PIGLET NUTRITION NOTES

Volume 1

PIGLET NUTRITION NOTES

Volume 1

Ioannis Mavromichalis, PhD

First published 2016

Published by
5M Publishing Ltd,
Benchmark House,
8 Smithy Wood Drive,
Sheffield, S35 1QN, UK
Tel: +44 (0) 1234 81 81 80
www.5mpublishing.com

A Catalogue record for this book is available from the British Library

ISBN 978-1-910455-62-3

Book layout by Servis Filmsetting Ltd, Stockport, Cheshire
Printed by Bell & Bain, Scotland
Cover photo by Mengtianhan | Dreamstime.com

About these notes

With the advent of Internet, there has been an increasing gap in the traditional way of disseminating information: that of a printed book. It is thus the aim of this series of booklets to provide a 'tangible' means of communicating practical, commercial, applied information to the global pig industry.

Each volume in the series will be a collection of timely, yet classic in being perpetual questions, topics on piglet nutrition, based on my education and experiences as a consulting nutritionist working on an international scale.

I am a strong believer in knowledge dissemination. But, in my efforts, I recognize I will probably not be correct at all times. At any given time, however, the information I present will always be what I believe to be true. At least, I will never recommend something that I am not prepared to use in my own work. And, for that, I have included a section at the end of each chapter called 'Personal Experience'. I believe you will find it interesting.

Expect the next volume to be published soon!

Ioannis Mavromichalis

Why limited references?

It is customary in scientific works of this nature to include a plethora of literature references to direct readers to further information and to substantiate all information provided.

I have opted not to follow this tradition, first because such references are nowadays freely accessible through the services of search engines in the Internet.

Secondly, this work makes no attempt of being a comprehensive review of any topic discussed; instead it is a compilation of research and personal experiences, aiming to support the commercial nutrition professional. Please, consider each chapter the equivalent of a large article found in any technical magazine.

Thirdly, in my experience, in-text references break the reading process and reduce the dissemination of knowledge as the non-research-oriented reader becomes distracted.

Nevertheless, I have included suitable references under tables when material was adapted from specific publications.

At the end, I would like to thank and collectively credit all those researchers whose work has enabled me to produce this work.

To my wife, Efi, for always being there for me . . .

Table of Contents

CHAPTER I

Protein nutrition guidelines

Protein, which is composed of amino acids, is the second most expensive nutrient in piglet diets (following energy and followed by phosphorus). Amino acids are used as building blocks for protein deposition, which is excessively rapid with modern fast-growing pig genetics. In addition, certain amino acids have other physiological roles. For example, tryptophan is involved in the regulation of feed intake, whereas glutamine is involved in gut immunity and cellular functions.

There are ten essential amino acids that cannot be synthesized by other amino acids in the diet. Among these essential amino acids, the most likely to become limiting in low-protein piglet diets are lysine, methionine, threonine, tryptophan, isoleucine, and valine. In contrast, the remaining four essential amino acids (histidine, phenylalanine, leucine, and arginine) are not problematic with most practical diets. This is why most feed formulation programs focus on the first three to five most limiting amino acids. Glutamine (a non-essential amino acid) supplementation in post-weaning diets may be also warranted on grounds of gut health, but no specific requirement has been established.

Feed-grade forms of lysine, methionine, threonine, tryptophan, and valine are commercially available, although cost for the last two is almost prohibiting for all but the most complex piglet diets. As a general rule of thumb, it is recommended to limit individual additions of feed-grade amino acids (especially lysine, which is first limiting) to 0.5% in piglet diets to prevent unbalancing the ideal protein profile, especially for non-essential and essential amino acids that are not included in diet formulation. In reality, however, up to 0.75% is added in antibiotic-free diets, which must be very low in crude protein to avoid digestive upsets.

Lysine

Pigs require lysine for maintenance and protein deposition in lean tissue gain. For maintenance, 36 mg true ileal digestible lysine are required daily per kilogram metabolic weight ($BW^{0.75}$). For protein deposition, 120 mg true ileal digestible lysine are needed per gram protein, reflecting the average concentration of lysine in muscle protein (70 mg/g) and the marginal efficiency (58%) of converting dietary true ileal digestible lysine to muscle protein-bound lysine.

Assuming that weight gain in piglets is 16% protein, a factorial approach may be used in determining requirements and dietary specifications with known feed intake levels. For example, a 10-kg pig growing 400 g/day on 600 g feed requires 0.2 and 7.7 g digestible lysine for maintenance and protein deposition, respectively. Therefore, a diet with at least 1.3% digestible lysine (without allowing for any margins of safety) would be needed to cover the daily requirement for lysine. Had actual feed intake been 750 g/day, dietary digestible lysine concentration would need to be only 1.05%, a remarkable 20% reduction! This example illustrates the importance of appropriate estimates of actual feed intake before setting dietary lysine specifications for a specific farm or group of pigs.

Factors that prevent pigs from attaining their genetic potential for growth also affect lysine requirement, because (1) maintenance requirement is trivial (even at 30 kg body weight it is only 0.5 g/day), and (2) protein deposition in young pigs increases linearly with increasing weight gain and feed intake. Health status is perhaps the most important determinant of growth during the nursery phase, because piglets of high health status have greater biological capacity for protein deposition.

In a classic study at Iowa State University, it was demonstrated that healthy pigs, such as those reared off-site under segregated early weaning management protocols, required 25% more lysine to maximize performance compared to challenged pigs that did not benefit from increasing dietary lysine concentrations. Interestingly, the efficiency of lysine utilization for protein accretion between these two groups of pigs was similar. In practical terms, when pigs reduce feed intake because of sub-optimal health, it is of no use to increase dietary lysine specifications.

For feed formulation purposes, dietary lysine specifications should be determined based on the lysine to energy ratios, examples of which are in Table 1.1. These values include a moderate margin of safety and should allow for optimal growth performance under good management conditions.

Ideal protein profile

An evaluation of dose-titration studies, and composition of sow's milk and body tissue, reveals that pigs require amino

3

Table 1.1 *Recommended dietary true ileal digestible lysine to energy ratios (g/MJ) for piglet diets*[1]

Body weight (kg)[2]	Lys:DE[3]	Lys:ME[3]	Lys:NE[3]
4–6	0.90	0.95	1.25
6–8	0.85	0.90	1.20
8–12	0.80	0.85	1.15
12–20	0.75	0.80	1.10
20–30	0.70	0.75	1.05

[1] One MJ equals 239 Kcal.
[2] For other weight ranges use average of adjacent values.
[3] DE = digestible energy, ME = metabolizable energy, NE = net energy.

acids in a specific balance or pattern (ideal protein profile) that remains constant regardless of the magnitude of absolute requirements. Although the amino acid balance is different for maintenance and growth, a single profile is satisfactory for piglets because maintenance needs represent a rather small fraction of total requirements up to 30 kg body weight. The ideal protein profile is not affected by gender, genetics, dietary energy density, feed intake, rate or composition of gain, and environmental conditions that determine, nonetheless, absolute requirements.

The ideal protein profile concept relates the requirement for all amino acids to a reference amino acid for easier diet formulation. In pigs, lysine is the most common reference amino acid because:

- it is the first-limiting amino acid in practical diets,
- its requirement is better established than requirements for other amino acids,
- it is used exclusively for protein synthesis, and
- its chemical analysis is relatively simple.

Table 1.2   *Ideal amino acid profiles for piglet diets*

Amino acids	Commercial	Sow's milk	USA (NRC, 1998)	France (ITP, 2002)	UK (BSAS, 2003)
Lysine	100	100	100	100	100
Methionine[1]	30	33	26	30	30
Methionine+ cysteine	60	56	56	60	59
Threonine	65	55	64	65	65
Tryptophan	19	16	18	19	19
Isoleucine	58	55	54	–	58
Valine	70	73	68	–	70

[1] Methionine can cover cysteine requirements with an efficiency of 81%.

Diet formulation using an ideal protein profile requires first the establishment of a lysine specification, which is available through dose-titration studies (or a factorial assembly process using maintenance and growth needs, plus a safety margin). Then, dietary specifications for other essential amino acids are calculated based on the ideal protein profile (Table 1.2). In this way, the need for dose-titration studies for each amino acid for every possible weight range and condition is eliminated.

For example, a diet for 10-kg piglets would require 1.15% digestible lysine at 10 MJ NE (Tables 1.1 and 1.2) and 0.22% digestible tryptophan (1.15 × 0.19) based on a 19% tryptophan to lysine ratio. Likewise, a similar diet for 25-kg pigs would require 1.05% digestible lysine and 0.2% digestible tryptophan (1.1 × 0.19). Tryptophan is used here as an example for all other amino acids, if only because it is extremely expensive as a feed-grade amino acid supplement. It is evident from this example that, even though pigs differ in body weight and require different levels of dietary lysine and tryptophan, the ratio between the two amino acids remains constant at 19%. The ideal protein profile concept is of paramount importance in formulating low-protein diets with increasing concentrations of feed-grade amino acids as it maintains the balance among all amino acids.

The ideal protein concept was originally developed based on the theory that providing amino acids closer to the actual requirements would enhance performance because excess amino acids would not have to be catabolized at an energy cost for the animal. Yet, there is no scientific evidence to support that pigs fed practical diets cannot handle slight excesses of amino acids, which are unavoidable when diets are balanced on the first limiting amino acid. Growth performance, however, is affected when an amino acid is supplied in toxic concentrations or in excess of its requirement when a more limiting amino acid is deficient. For example, in an older study, growth depression caused by 0.2% added methionine was alleviated only by the addition of 0.2% lysine, which was first limiting.

Limiting amino acids

Because all sources of amino acids (protein) are expensive, dietary excesses are always against profitability. The amino acid present in the least amount relative to its specification is referred to as the first-limiting amino acid and dictates performance. When the specification for the first-limiting amino acid is met, the next amino acid with the least concentration relative to its specification becomes second-limiting and, in turn, it dictates performance.

The order by which each amino acid becomes limiting depends entirely on ingredient selection and adopted ideal protein profile. Thus, isoleucine becomes limiting in diets rich in blood products, histidine in diets based heavily on barley, and tryptophan in diets based on maize and meat meal. In such diets, supplementation with specific feed-grade amino acid(s) and (or) higher levels of protein-rich ingredients becomes essential to meet specifications. It is therefore suggested that specifications for all ten essential amino acids are included in diet formulation to eliminate the risk of potential deficiencies when using novel ingredients. However, this might be impractical and this is why most formulation software includes the first six limiting amino acids only.

In most practical piglet diets, lysine is clearly the first-limiting amino acid. In low-protein diets, however, other amino acids become rapidly limiting as protein level decreases. When dietary protein concentration is decreased by 2 to 4 percentage points, methionine, threonine, and tryptophan are most likely to become limiting, with valine and isoleucine becoming limiting with more severe reductions in crude protein concentration.

Low-protein diets

Reduction of dietary protein in nursery diets is currently deemed beneficial for improving animal health and reducing environmental pollution. In addition, the advent of antibiotic-free nutrition dictates the use of very low protein diets to avoid supplying excess protein to undesirable pathogens endemic to large intestine microflora.

In general, low-protein diets are formulated to meet amino acid specifications with the least amount of crude protein. Such diets are naturally more digestible than conventional diets because feed-grade amino acids (100% digestible) and (or) protein sources of higher biological value (such as blood products, fish meal, refined vegetable protein sources) are used in greater concentrations. In result, less undigested protein remains for bacterial proliferation and excretion to the environment.

It has been shown since the 80s that piglets fed diets low in crude protein are more resistant to digestive upsets from *Escherichia coli* infections. Low-protein diets have been shown also to reduce nitrogen excretion by 30 to 50%, depending on original specifications, ingredient selection, and magnitude of reduction in dietary protein concentration. In general, for each percentage unit reduction in dietary crude protein, nitrogen excretion is expected to be lower by approximately 8%. Suggested levels for crude protein are presented in Table 1.3.

As dietary protein concentration is severely reduced, several unique concerns arise in diet formulation. First, the balance between essential and non-essential amino acids is disturbed. This ratio should be around 50:50 to allow optimal

Table 1.3 *Suggested levels for crude protein in commercial piglet diets*

Body weight (kg)	NRC (1988)[1]	Commercial[2]	Antibiotic-free[3]
4–6	26	24	19
6–8	24	22	18
8–12	22	20	17
12–20	20	18	16
20–30	18	16	15

[1] Only values from the 1988 publication are available, and they reflect the heavy use of growth-promoting antibiotics.
[2] Average values reflecting a wide range of feeds, worldwide.
[3] Even lower levels are possible with the use of more feed-grade amino acids, but growth performance might be affected.

nitrogen retention and utilization. Second, the importance of predicting lysine requirement increases because the more expensive feed-grade amino acids (tryptophan, valine) supply a larger fraction of the total. Third, it is recommended the net energy system be used in formulating low-protein diets based on true ileal amino acid digestibilities. Fourth, amino acid antagonisms may develop if arginine becomes excessive compared to the lysine concentration or leucine compared to the sum of isoleucine and valine concentrations. Fifth, sodium bicarbonate additions may be needed to buffer the acidic effect of excessive amounts of free amino acids in diets past the post-weaning period.

Taking into account benefits and challenges of a low-protein nutrition approach for piglets, it appears that a moderate reduction (2%) in dietary protein is quite feasible. A more drastic reduction (4%) requires a higher degree of accuracy in diet formulation, and perhaps changes in ingredient selection and nutrient matrices. Reducing protein beyond 4% is expected to reduce performance in commercial production systems, but in most of these cases, growth performance is a concern secondary to avoiding digestive upsets in the post-weaning period.

Tryptophan as a functional nutrient

The role of tryptophan lies beyond protein synthesis to include biological functions such as appetite regulation and control of sleep via its metabolite serotonin. Because tryptophan is rather expensive in its feed-grade form (up to 10 times more expensive than other feed-grade amino acids), an accurate estimate of tryptophan requirement is essential in balancing growth performance and cost of production. European researchers evaluated the tryptophan requirement of pigs between 10 and 30 kg body weight using a basal diet formulated with cereals, peas, whey, and fishmeal. This basal diet was then fortified with increasing doses of feed-grade tryptophan to achieve tryptophan to lysine ratios between 15 and 25%. Results indicated that growth rate and feed efficiency were maximized at a ratio between 19 and 21% for pigs between 10 and 20 kg body weight, whereas growth from 10 to 30 kg was optimized at about 19%. These results are also supported by Canadian researchers, who found the digestible tryptophan and lysine requirements for protein deposition to be 16.8 and 89.8 mg/g, respectively, giving a ratio of tryptophan to lysine of 19%. In contrast, multiple researchers have indicated that adding up to 24% tryptophan (as a ratio to lysine) is essential when pigs are faced with immunity challenges and (or) as a way of increasing feed intake through the appetite regulation pathway. At the end, it is left to the discretion of the nutritionist to decide which tryptophan ratio to select based on expected outcome.

Expressing requirements and specifications

Typically, requirements and dietary specifications are both expressed in relative terms as percentages (%) or amount per kilogram (for example, g/kg) feed, because references to daily requirements are not practical outside the sphere of research or academic discussion.

Most nutrient requirements have been determined on a total nutrient basis. The term "total" refers to the requirement of a nutrient that can be determined without taking into account any interactions with the animal organism. This is acceptable as long as practical formulations do not deviate much from those used in the original experiments – a rare finding under commercial conditions.

Expressing requirements and dietary specifications on a digestible basis (total minus nutrients excreted in feces) is strongly recommended to ensure adequate supply of nutrients for maximal performance. Unfortunately, existing digestibility coefficients for most ingredients are derived from studies with growing-finishing pigs, and thus their value in formulating diets for young pigs is at best questionable. Plus, digestibility coefficients are not available for all ingredients used today in piglets. Nevertheless, today most nutritionists use digestibility values for amino acids on a constant basis.

For amino acids, the use of terminal ileum digestibility values are clearly more accurate than fecal digestibility values, because the latter include bacterial protein synthesis in the large intestine that contributes little if anything to the protein nutrition of the animal. A step further, deduction of endogenous amino acid losses yield true ileal digestibility, which is of course more representative than apparent ileal digestibility. Apparent ileal amino acid digestibility values tend to underestimate low-protein ingredients, as they tend to increase with increasing dietary protein concentration.

For most commercial piglet diets, the following rules of thumb from NRC (1998) may be used to navigate between amino acid digestibility values.

- from total to true ileal, multiply by 0.90,
- from total to apparent ileal, multiply by 0.82, and
- from true ileal to apparent ileal, multiply by 0.94.

Practical safety margins

Most nutrient requirement studies have been conducted under experimental conditions that would be deemed more favorable than current commercial conditions. In such experiments, diets are carefully prepared from ingredients of known chemical composition, and they are fed shortly after mixing to avoid nutrient losses. Thus, most commercial nutritionists add a margin of safety over published nutrient requirement estimates, based on experience and limited experimental work, to account for variables that are hard or impossible to control or even predict during feed preparation and use. The following is a non-exclusive list of cases that call for margins of safety.

Label guarantees. By definition, in least-cost formulation the probability for any nutrient to be below target is always 50%. Thus, formulas may be slightly over-fortified to meet declared concentrations in feed labels as required by law.

Nutrient variability. Certain ingredients can be highly variable in terms of nutrient composition. Agro-industrial products of unknown quality and origin can be highly problematic if they are not assayed frequently, especially if they make up a substantial part of a diet.

Formulation basis. Diets formulated on total rather than digestible nutrient basis are often over-fortified to account for unknown variability in nutrient digestibility, especially in low-quality and novel ingredients.

Nutrient losses. Several nutrients, and especially amino acids, are susceptible to destruction during thermal processing (pelleting, expanding, flaking), and also during prolonged storage, where significant losses may occur due to the Maillard reaction process.

The "average" pig. Published nutrient requirements usually refer to the "average" pig, which in reality does not exist. This definition of convenience leaves half the pigs in a group undernourished while the other half are overfed. Production systems that place emphasis on rapid turnover require high rates of growth that can only be sustained by diet over-fortification to allow for proper nutrition of most pigs. This is a very exciting topic that brings together feed formulation, statistics, and growth modeling software.

Minimizing unknown factors and monitoring quality in ingredients and finished diets obviously reduces the need for excessive safety margins. If animal performance (weight gain and feed intake) is known from records, then a feeding program with minimal safety margins that closely meets requirements can be easily designed by a qualified nutritionist. Nevertheless, it is always beneficial to challenge dietary specifications in specific pig production systems that manage to monitor performance. To this end, margins may be gradually increased (or decreased) in 5 to 10% increments and animal performance be evaluated. This exercise should cover a long-term period and a wide range of conditions.

Typical margins of safety can range from 5 to 30%, depending on the gravity of the factors taken into consideration. As a rule of thumb, a 10% safety margin over requirements is adequate

Table 1.4 *Recommended margins of safety (%) for dietary lysine specifications in piglet diets*

Criteria	With quality control	Without quality control
Minimal feed storage		
Pellets	10	15
Meal	5	10
Prolonged feed storage		
Pellets	15	20
Meal	10	15

when a quality-control program closely monitors incoming ingredients and finished products. Higher safety margins are needed when significant variations in nutrient composition are expected due to ingredient selection, processing conditions, storage, and animal performance (Table 1.4).

But not all nutrients require the same safety margins. For example, if a considerable fraction of the variability in lysine concentration is due to destruction by the Maillard reaction, then lower safety margins can be justified for the rest of the amino acids that are more stable during thermal processing. This can be easily achieved (1) by using a "dummy" lysine dietary specification to estimate other amino acids based on an ideal protein profile, or (2) by calculating specifications based on an ideal protein profile and then applying different safety margins separately to each amino acid.

Personal experiences

The basis of my PhD thesis was on determining the fifth-limiting amino acid in corn-soybean meal-based diets. It was valine.

I have also conducted studies to establish the requirement of this amino acid in relation to lysine: 72%. These findings have now been verified by numerous researchers, and based on all this work, the use of feed-grade valine has become possible.

Since then, I have realized the importance of low-protein diets when antibiotics are excluded from piglet diets. In fact, this exercise has enabled me to "test" under commercial conditions the use of low-protein diets even when growth-promoting antibiotics were used. At the end, I realized that piglets always do better when excess dietary protein is removed from their diets.

I have always used the net energy system and true ileal digestible amino acids for formulation purposes. It made my work relevant to my international clientele and I had no worries when I applied the same principles whether I was formulating a corn-soybean meal diet in the USA or a diet with 20 different and most of them "exotic" ingredients for a client in The Netherlands.

For that, I have relied on the net energy system from France and the digestible amino acid values from the USA. I find these two sources as accurate as any, as they are based on considerable and groundbreaking research work. Other systems have been evaluated but they failed to produce better outcomes as they too depended on the original work conducted in the sources described above.

Finally, my work in Asia has taught me that low-protein diets might not be always acceptable; marketing and tradition still have a role to play in commercial nutrition. To this end, I have been involved in finding ways to minimize the negative effect of excess crude protein by using other ingredients, but that's material for a future chapter.

Notes on protein

CHAPTER 2

Antioxidant nutrition: vitamins E and C

Antioxidant nutrition presents a paradox: despite its immense recognition among human nutritionists, those of us involved with animal nutrition largely ignore this topic. It is perhaps because antioxidant nutrition has been associated with aging – not a major problem for farm animals, but, nevertheless, oxidative stress is common in farm animals consuming average-quality feeds. So far, adding a feed antioxidant, such as the traditional ethoxyquin, is considered the end of the matter for most nutritionists; to the contrary, new evidence suggests that feed protection is only the beginning. Indeed, oxidation occurs even within the organism where products like in-feed common antioxidants cannot reach and protect. To this end, vitamins such as E and C have a significant role to play; after all they are considered as super-antioxidants in human nutrition circles. Notwithstanding their effectiveness, these two vitamins remain expensive, and as such they must be employed at the right amount and occasion if they are to be of benefit to farm animals.

Oxidative stress

The normal process of metabolizing nutrients invariably leads to production of undesirable compounds, among which, perhaps the most dangerous, are those called free-radicals. These chemical compounds "steal" electrons from nearby molecules. In doing so, they harm cells through long-term irreparable damage. For example, it has been shown that cellular membranes and DNA (two very important components in all cells) are particularly vulnerable to free-radical damage. Such damage, especially to cellular membrane, can lead to severe neurological damage, whereas DNA damage causes malfunction of the genetic apparatus. It is believed that aging is due (at least partly) to the natural decline in controlling free-radical damage.

The damaging effect of free-radicals is neither acute nor immediate. It is a slow but steady long-term process that requires the build-up of damage before health starts to deteriorate and certain organs or tissues begin to malfunction. Or, as is quite usual in the case of the immune system, the effects from excessive free-radical exposure do not become apparent before the weakened organism is infected. Perhaps, this problem in assessing the true extent of pre-existing free-radical damage has created some disappointment when it comes to control through dietary supplementation with natural antioxidants.

Oxidative stress can be defined as an imbalance between free-radicals and antioxidants at cellular level. In animals, high oxidative stress is due mainly to improper nutrition (high lipid content, mycotoxin contamination). In addition, thermal stress and several diseases can cause additional oxidative stress. It has been suggested that even the process of vaccination can cause oxidative stress.

Natural antioxidants

Antioxidants are compounds that neutralize scavenging free-radicals. The main mode of action of antioxidants is by donating hydrogen atoms from hydroxyl groups to terminate the free-radical oxidation of biomolecules. In essence, they stop oxidation.

Naturally, the organism is not defenseless against free-radicals. There is a great number of innate defense mechanisms against free-radicals, among which are compounds such as superoxide dismutase, glutathione peroxidase, and Q10 – all produced by the animal under normal conditions. These antioxidants also work by simply donating electrons to scavenging free-radicals, rendering them virtually harmless. Here it should be noted that there are countless molecules that can donate electrons, but dedicated metabolic "antioxidants" are specific for certain processes working with greater efficacy and specificity – and they spare all other molecules required for other purposes. The need for a great number of dedicated antioxidants is due to the fact that antioxidants are not interchangeable, and this applies to those provided through feed ingredients.

Nature has ensured that animals receive daily a good amount of extra antioxidants through their feed. Such antioxidants include familiar names, such as vitamin C, vitamin E, and selenium. But there are more such compounds, currently not so well known. These include beta-carotene (pro-vitamin A) and several phenols belonging to the families of flavonoids and tannins, which abound in seeds and fruits. Even wood contains several antioxidant compounds. Clearly, there is no lack of antioxidants when it comes to nature, which should illustrate their importance in sustaining life in all living organisms, and again their lack of interchangeability – something that could help reduce the confusion in interpreting the results of many experimental studies. In fact, currently there is a great interest in obtaining antioxidants from such materials as grape seeds, pine trees, and chestnuts, to mention only a few.

Measuring oxidative stress

It is quite disconcerting that currently there is no accurate and practical assay to quantify the extent of free-radical damage in domestic animals. Likewise, there is no quantification of antioxidant requirements based on species, age, condition, health, etc. So far, all attempts towards supportive antioxidant nutrition have been on a qualitative-empirical basis, with highly variable results. Thus, there is a good number of peer-reviewed published research reports in which the positive effects of certain antioxidants is clearly illustrated; such is the case with vitamins C and E, and to some extent with selenium. At the same time, there is an equal number of similar reports for the same antioxidants where such effects were not observed. Perhaps the confusion is in the difficulty of identifying conditions under which animals have an increased need for antioxidants, and connecting this need not only with the correct dosage of a supplemental antioxidant, but also with the correct form of antioxidant.

This enigma is perhaps illustrated most clearly in the case of vitamin E. Today, there is a great number of veterinarians and nutritionists who recommend extremely high doses of vitamin E in piglet feeds. They usually employ such measures when piglets are under any form of stress. Such cases will be, of course, those of weaning, and as such, many diets contain three to five times the normal requirement for this vitamin. In addition, persistent diseases that do not retreat despite common cure practices have been resolved with injections of vitamin E (along the normal medical treatments). Naturally, piglets do not require such high levels of vitamin E when raised under normal conditions. But the unexpected stress perhaps raises their need for supplemental antioxidants – and vitamin E is powerful enough. Interestingly, vitamin C does not work in a similar way, emphasizing the uniqueness of each antioxidant. Instead, vitamin C has been suggested as a means of combating heat stress, which also leads to oxidative stress.

Apparently, the response to antioxidants is related to their quantitative requirement, which appears to increase during conditions of stress. It is unclear whether this is due mainly to a reduced efficacy of internal antioxidant systems or due to an increased production of free-radicals. As is the case for most natural systems, it is perhaps a combination of both causes, complicated by the type of the stress factor – not to mention other complicating factors such as animal base health, condition, age, productivity, genetics, etc.

Today, there is considerable debate among nutritionists whether super-intensive productivity, which requires an increased feed intake and enhanced nutrient metabolism, causes damaging oxidative stress in modern animals. The debate is not whether increased metabolism results in higher generation of free-radicals: this cannot be denied. Rather, the debate is whether this unavoidable increase in free-radical production can cause enough oxidative stress to negatively affect animal productivity and health.

Naturally, the notion that super-intensive productivity is accompanied by a damaging oxidative stress requires no further explanation. It is interesting, however, to briefly present some counter-arguments, which might help explain why in some research reports antioxidant supplements appear to be of little benefit. First, young animals are more resistant to oxidative stress compared to older animals that suffer from age-related reduction of efficacy in all systems. Second, certain animals that experience super-intensive growth patterns, such as growing pigs, have limited lifespans. As mentioned above, the damaging effects of free-radicals require time to fully manifest. Third, with a higher feed intake there is a higher intake of natural antioxidants, already present in most feed ingredients. Fourth, commercial feeds already contain enough safety margins for the most common antioxidants, such as vitamin E and selenium.

Vitamin E in piglet feeds

Although vitamin E belongs along with vitamins A, D and K to the fat-soluble vitamins, it is actually quite unique in being also a powerful antioxidant.

Vitamin E is a natural compound found in two forms: tocopherol and tocotrienol. Each form is encountered as alpha, beta, gamma and delta tocopherol, and tocotrienol, respectively. Of all these, d-alpha-tocopherol has the greatest biological efficacy, followed by the l-isomer, whereas the dl- form is usually taken as the standard in discussing vitamin E nutrition. For example, the International Unit (IU) of vitamin E (a common measure of vitamin E potency and concentration in feeds) is taken as the activity of 1 mg of dl-alpha-tocopherol acetate (Table 2.1).

Sources. The primary source material for vitamin E has been the tocopherols found in leguminous seeds high in oil, mostly

Table 2.1 *Conversion factors for common vitamin E forms*

---

1 mg = 1.00 IU dl-α-tocopheryl acetate
1 mg = 1.36 IU d-α-tocopheryl acetate
1 mg = 1.10 IU dl-α-tocopherol
1 mg = 1.49 IU d-α-tocopherol

NRC (2012), Nutrient Requirements of Swine, USA.

soybeans. Nevertheless, such natural sources are prone to rapid oxidation. As such, modern practices call for supplemental vitamin E to be provided through a vitamin premix. Such premixes almost invariably contain enough vitamin E to cover completely the entire vitamin E needs of the animal, disregarding thus any vitamin E provided by natural ingredients. In addition, these premixes often contain surplus vitamin E to safeguard against potential losses during feed processing and storage, and from reaction with other nutrients.

Requirements and specifications. Minimal requirements for prevention of deficiency symptoms and adequate performance under adequate conditions for relatively healthy animals have been established with great accuracy by the NRC (2012; Table 2.2). It should be mentioned here that the NRC values have not changed since the previous edition (1998) and even those values were based in trials several decades old. Lamentably, those trials remain the only valid ones as more recent efforts have failed to meet the strict scientific standards required to make them credible enough to challenge existing knowledge.

Nevertheless, commercial nutritionists always increase basic requirement values with various safety margins to derive at commercial dietary specifications. Such an effort combining research data with commercial experiences was undertaken by a scientific group of the British Society of Animal Science (BSAS 2003). These values are significantly higher than the NRC (2012) minimal requirements, but still quite lower than recommendations by several vitamin suppliers.

Another source of vitamin E practical recommendations is the National Swine Nutrition Guide (USA, 2010). These values are not much different than the BSAS (2003) specifications, and represent common university thinking today in the USA.

Finally, based on commercial experience, maximal and average dietary specifications for piglets less than and more than 10 kg body weight are presented in Table 2.2. The higher specifications are used in high-quality, and often expensive, piglet feeds for recently weaned piglets, or for milk replacers and creep feeds.

Deficiency. Vitamin E, being a primal antioxidant, is required in a myriad of functions within the body on a daily basis. As such, it is an extremely important vitamin for major functions such as growth and development, respiration, anabolism and catabolism. Thus, even a slight deficiency that may not cause any clinical signs can result in reduced performance and thus

Table 2.2   *Vitamin E requirements and commercial practice (mg/kg)*

		Piglet body weight (kg)	
Component	Year	<10	>10
National Research Council, USA	2012	16	11
National Swine Nutrition Guide, USA	2010	60	30
British Society of Animal Science	2003	80	50
Maximal commercial practice		250	100
Average commercial practice		100	50

profitability. As most commercial feeds are over-fortified with vitamin E, most cases of deficiency are associated with (a) mixing errors, (b) reduced vitamin E potency in the feed or premix, or (c) unexpected increase in antioxidant requirements due to stress factors (nutrition, environment, health, management).

Classical symptoms of deficiency include skeletal and cardiac muscle degeneration, gastric parakeratosis and ulceration, anemia and liver necrosis, and yellow discoloration of lipid tissues. Vitamin E deficiency might also cause sudden death, and it has been implicated in leg problems including arthritis and swollen joints.

Losses, natural and synthetic. Vitamin E concentration in natural feedstuffs is variable and it is also affected by manufacturing processes. Being fat-soluble, vitamin E is associated with the lipid fraction of ingredients, such full-fat soybeans and wheat germ. Fine milling of such feedstuffs will expose their lipids to environmental oxidation, which in turn will consume natural occurring vitamin E. Furthermore, vitamin E, and supplemental vitamin E in particular, is very sensitive to organic acids. As the latter become a mainstream ingredient in piglet diets, it is perhaps important to reevaluate the levels of vitamin E in such diets to prevent deficiencies. Of course, adding already oxidized lipid sources in piglet feeds is a sure way to cause unnecessary vitamin E losses even before the feed is consumed.

Natural antioxidants versus vitamin E. Naturally occurring antioxidants, such as certain polyphenols, are now available as extracts or part of certain specialty ingredients. These compounds exert a strong antioxidant effect that can spare or replace the role of vitamin E. Indeed, at times when vitamin

E becomes extremely expensive, such additives find increased usage. There is a replacement factor (for example, x mg of such polyphenols from this source equal y IU vitamin E), but this is different for each product as not all polyphenols are the same (in fact, it is one huge and widely diverse group of thousands of compounds). In addition, research is still ongoing and remains unclear, but promising. In general, such products can replace at least a part of vitamin E, especially when this vitamin in used at supra-nutritional levels.

Vitamin C in piglet feeds or water

Vitamin C (also known as ascorbic acid or ascorbate) is a naturally occurring metabolite of glucose. Most species, including pigs, are able to synthesize vitamin C. It is only the absence of the enzyme L-gulonolactone oxidase that makes vitamin C a dietary essential nutrient for humans and such animals as primates and guinea pigs. Nevertheless, there is sufficient evidence today to suggest pigs may benefit from vitamin C supplementation under certain conditions, but the results are still unclear.

Biological role. The metabolic functions of vitamin C are many and important. They include free-radical detoxification, vitamin E rejuvenation, collagen, norepinephrine and carnitine biosynthesis, and neurotransmitter and metal ion metabolism. Also, vitamin C plays a significant role in the functionality of the immune system. During periods of stress (for example, early weaning, excessive heat, disease challenge, fighting, crowding) body stores of vitamin C are depleted quickly and high concentrations of vitamin C metabolites are excreted in urine.

This latter phenomenon has led many scientists to believe supplemental doses of vitamin C may be warranted under such adverse conditions.

From a biological point of view, pigs are able to synthesize vitamin C as early as one week old. Moreover, ample quantities of vitamin C are supplied via the sow's colostrum and milk, assuming piglets consume enough milk. However, early weaning at an age of 14 to 28 days imposes an immense nutritional, social, and environmental stress on pigs. In fact, the plasma concentration of vitamin C drops drastically in pigs shortly after weaning. This has been taken as evidence of either an insufficient biosynthesis rate during the period of stress, or an increased requirement for vitamin C during the same period. It is most likely that both factors combine to create an apparent vitamin C deficiency during weaning time. Therefore, it has been proposed that nursery pigs might benefit from dietary supplementation with vitamin C.

Early studies. In an early study (Table 2.3), pigs weaned at 4 and 5 weeks of age were offered diets fortified with increasing doses of vitamin C (0, 330, 660, and 990 mg/kg) in a 28-day growth assay. Overall, weight gain was improved by 17% and feed intake by 13% in pigs offered the vitamin C-fortified diets. There was no effect on feed efficiency. This was an exemplary trial and sparked a wave of enthusiasm among nutritionists followed by many follow-up trials on vitamin C.

Unfortunately, a great number of experiments conducted in several North American universities did not succeed in repeating the above positive results. Fortification of piglet diets with vitamin C at various levels ranging from as little as 75 to as much as 990 mg/kg did not result in enhanced

Table 2.3 *Studies with weaned pigs given vitamin C supplementation*

Reference	Treatments/trials
Yen and Pond, 1981	0, 330, 660, 990
Mahan and Saif, 1983	350 mg/kg or injection = 40 mg/2d
Mahan and Saif, 1983	(0, 450, or 900 mg/kg) × (± antibiotic)
Mahan and Saif, 1983	(0 or 900 mg/kg) × (0 or 60 mg/kg Cu)
Nakano, et al., 1983	0, 350, 700 mg/kg
Yen and Pond, 1984	(0 or 660 mg/kg) × (± antibiotic)
Kornegay et al., 1986	(0 or 700 mg/kg) × (19 vs 27° C)
Chiang et al., 1985	0, 150, 300, 450 mg/kg
Yen and Pond, 1987	(0 or 660 mg/kg) × (.13 or .25 m^2/pig)
NCR-89 Committee, 1989	(0 or 625 mg/kg) × (.124 vs .248 m^2/pig)
Mahan et al., 1994	0, 50 or 500 mg/kg (stable vitamin C)
de Rodas et al., 1998	0, 75, 150 mg/kg (stable vitamin C)

Except for the first, all other trials did not find any positive response to vitamin C supplementation in piglets post-weaning.

Studies published in the Journal of Animal Science.

growth performance, hemoglobin concentration, immunity, or survivability. Also, injections with vitamin C (40 mg/kg every other day) did not enhance growth performance. The same lack of response was observed in pigs that were crowded or stressed by cold temperatures. In fact, in a large study involving 11 experimental stations and 1,296 nursery pigs, responses to dietary vitamin C supplementation (625 mg/kg) did not occur, even when available space was reduced to only 0.128 m^2 per pig. Also, vitamin C fortification failed to elicit any growth performance responses in piglets over and above those obtained from antibiotics or copper sulphate supplementation.

Possible reasons for failure. Vitamin C is sensitive to a variety of environmental factors including heat, oxygen, and alkalinity. It has been proposed that the lack of response observed in these experiments was due to inevitable vitamin C losses that occurred from manufacturing, and during storage and feeding. However, in most of the experiments, the researchers were aware of this problem and they tried to avoid considerable vitamin losses by frequent preparation and even cool storage of the diets. Nevertheless, some considerable losses were not possible to avoid, given the sensitive nature of this vitamin.

It has been suggested that the plasma concentration of vitamin C in pigs is not a good indicator of their inability to synthesize vitamin C or of their increased demands. The drop in plasma vitamin C in recently weaned pigs can not be taken as evidence of a decrease in biosynthetic rate, because this decrease is likely caused by the fact that the pigs no longer have access to sow's milk. Piglets are able to synthesize vitamin C from birth, and at periods of high requirements they are capable of adjusting biosynthetic rates to match demands. At the same time, it is perplexing why sow's milk contains so much vitamin C to begin with.

Ongoing research. A good number of studies testing vitamin C in pigs, and especially piglets, are published almost each year. Some of these studies show a positive effect, some others no effect at all. Based on the available evidence, it is impossible to decide with any confidence whether vitamin C supplementation is beneficial or not. Interestingly, in recent studies, even the use of more stable forms of vitamin C provided results that were sometimes positive and other times without any benefit.

What is really needed is to understand under which conditions vitamin C is beneficial. Clearly, we cannot use vitamin C as a routine supplement in all pig diets, like we do for example with vitamin E. But, certainly, there are enough indications that vitamin C can play a significant role. It is this last aspect that we currently don't understand.

NRC 2012. The latest review from the National Research Council in the USA (2012) states:

> *If a supplemental vitamin C need exists, it would seem to be of a transient need during times of stress when feed intake may be limited.*

Due to the lack of convincing evidence, this scientific body of merit did not venture in offering an estimate of such need, until further research clarifies the conditions under which vitamin C supplementation may be warranted.

Practical considerations. Despite the lack of a consensus, some nutritionists have already made up their mind regarding the use of vitamin C. They base their decision not only on available scientific data but also on their own practical experiences. Most nutritionists do not include vitamin C in piglet diets. Some will include it in the first diet post-weaning, as a precautionary measure. Others will include it in all piglet diets during summer months, when heat stress is expected to affect feed intake. Finally, some others will recommend using vitamin C as a water supplement during heatwaves, and certainly after weaning. As might be expected, levels used are as variable as opinions about vitamin C. The highest recommended dosage is 1000 mg/kg, but as little as 100 mg/kg is not uncommon, especially in less expensive piglet feeds.

Personal experiences

I have studied vitamin C since my graduate days but I have yet to use it routinely in the piglet diets I design; I am just not convinced. Having said that, I do believe that vitamin C supplementation has a role to play during heat stress.

When it comes to vitamin E, I have always opted for moderate levels, letting veterinarians decide when supra-nutritional levels might be needed to address certain persistent disease issues. On the other hand, I have always advised the use of the highest possible quality sources of oils in piglet feeds, not allowing rancidity to consume valuable vitamin E. Rancidity depresses feed intake, which for me is of primary importance.

I am currently exploring the whole topic of antioxidant nutrition, and I am looking into potential antioxidants beyond the two vitamins described in this chapter. More precisely, I am studying the use of phenolic compounds.

In all, I am just as perplexed as most pig nutritionists by the lack of attention paid to oxidative stress under intensive modern production methods. I expect a surge in use of antioxidants, once the scientific community manages to measure the damage caused by this problem and associates it with specific antioxidants. I consider antioxidant nutrition a topic with a great future, if only funding could be secured for trials.

Notes on vitamin E

Notes on vitamin C

CHAPTER 3

Immunoglobulins in piglet nutrition

All pig diets are formulated to meet specific requirements for energy, protein (and amino acids), fiber, vitamins, and minerals. In piglet diets, we also add dairy products as a source of lactose, and perhaps a number of vegetable- and (or) animal-derived protein sources to accommodate the limited digestive capacity after weaning. High-quality piglet diets will also contain a number of gut-microflora-controlling additives, such as antibiotics, organic acids, zinc oxide, etc. All these ingredients and nutrients are combined through a complex formulation process to produce a sophisticated piglet feed that meets the requirements of modern production: lack of diarrhea, high feed intake, rapid and efficient growth.

Yet, only a few decades ago, just before the last known vitamin (B12) was discovered, piglet diets required fortification with ingredients that had no apparent purpose or for reasons that we later found to be undiscovered nutrients in deficit. Examples abound: meat meal, a unique source of vitamin B12, was considered essential because this vitamin was discovered very late; dairy products were used not because piglets required lactose, but due to their high palatability. Even fishmeal was once considered essential, but this is no longer the case because now we know how to formulate on digestible amino acids.

The last such mystery ingredient was animal plasma, which gave rise to another indispensable "nutrient": immunoglobulins. Understanding how plasma works and how animals benefit from immunoglobulins are the next steps in mastering this nutrient.

Immunoglobulins

We have been providing piglets with immunoglobulins since we first started feeding them with dairy products, and specifically with whey powder-based diets. Indeed, a good-quality whey contains as much as 5% immunoglobulins (albeit of a rather weak nature). But, they still contribute significantly when whey is added at 10 to 20% in a piglet feed. Original research with whey demonstrated that the main effect of this ingredient was due to its high lactose content. Nevertheless, in these diets high levels of antibiotics were used, as was the norm at that age, and as such, any gut health effect was excluded from the resulting observations.

So, what are immunoglobulins? Simply, they are all-natural globular proteins produced by animals as part of their immune system. They are the last and most powerful line of defense against invading pathogens, such as viruses and bacteria. Their production is quite expensive for the organism and as such they are only produced when the pathogens are not cleared by other systems of defense.

Immunoglobulins are created specifically to target, in a lock-key model, the particular invading pathogen, virtually "tagging" it for destruction or elimination by other immune system components. Immunoglobulins recognize the target pathogen based on the latter's structure. It is not necessary to recognize the whole structure, as partial recognition is sufficient. This is one of the reasons why immunoglobulins produced by one species (in cow's milk, for example) work in other species (when fed to piglets, for example).

There are many classes of immunoglobulins, some of which are involved in allergic reactions, others have a more local function, others are prevalent in milk and eggs, and yet others have a more widespread reach. It is only logical to expect that immunoglobulins created specifically for a certain pathogen are more potent than generic immunoglobulins, and as such there is a difference in their effect, depending on potency and concentration.

Immunoglobulin types. Mammals produce immunoglobulins G (IgG), whereas birds produce immunoglobulins Y (IgY). These are the major immunoglobulins circulating in blood, and consequently end up in colostrum and eggs. Mature milk contains high levels of immunoglobulins A (IgA), which are rather weak and largely nonspecific working at the gut epithelium level. In contrast, IgG and IgY are very pathogen-specific, IgY binding with higher strength and with better efficiency than IgG, because the egg is a smaller "package" than milk — being the only way to pass these powerful immunoglobulins from mother to offspring.

Immunity in young animals

Immunoglobulins are part of the immune system, but in young pigs, this system is not fully developed; hence, the high significance of colostrum, milk, and post-weaning nutrition and health to disease prevention. To fully appreciate the impediment of a still-developing immune system, it merits reminding of the basics of immunity in newborn and newly weaned pigs.

Newborn pigs are protected against diseases by specific immunoglobulins provided by their mother (passive immunity) through colostrum and milk. Shortly afterwards, they start producing their own immunoglobulins as they come into contact with pathogens (active immunity). Unlike humans, who receive maternal immunoglobulins in the womb during pregnancy, sows possess a multilayered placenta that prohibits the transfer of immunoglobulins to the fetuses. Thus, intake of colostrum shortly after birth is essential for survival. During the first 24 hours after birth, transfer of immunoglobulins through the intestinal wall is highly effective, but later on not only the concentration of immunoglobulins in the colostrum and milk decreases drastically, but also the permeability of the intestinal mucosa reaches virtually zero levels. The piglet therefore, after receiving a quick immunity "starter kit", must build up its own (active) immunity very quickly in order to survive.

The usual time of weaning for pigs raised commercially is between three and four weeks of age. Unfortunately, this time coincides with the time when passive immunity has waned to its lowest, whilst active immunity has just started to emerge. As such, the weaned piglet is most vulnerable to pathogens immediately after weaning. The same pathogens that already affected piglets before weaning, but without success, find at weaning the opportunity to cause disease. Such is the cause of several enteric-digestive diseases, and this is further complicated by the rough nature of most piglet feeds (as opposed to the refined nature of sow's milk that fully matches piglet requirements).

Cross-species immunity transfer

It is well known that for orphan or surplus piglets bovine colostrum can be used effectively as a replacement. Indeed, most farms keep sufficient supplies of bovine colostrum, either natural in frozen form, or as a dry product that requires restitution. In either case, this colostrum is crucial for piglet survivability. Apparently, immunoglobulins in bovine colostrum confer enough protection to piglets, even though these immunoglobulins were not produced by a sow or even at the same farm where the piglets are raised.

In addition, today it is well recognized and amply supported by research findings that bovine animal plasma is as effective as porcine plasma. In fact, when porcine plasma was temporarily considered unsafe (perhaps unjustly) for use due to the spread of the porcine epidemic diarrhea virus in the USA and Canada, many feed manufacturers switched to bovine plasma; and apparently they did not miss anything in piglet performance. At the end, it was demonstrated that quality was more important than species source, as properly processed porcine plasma was found to be safe enough. Nevertheless, this is one more case of effective cross-species transfer of immunity.

One final example comes from very early times, when different types of farm animals were kept together in a farm. There, free-ranging hens were virtually in contact with all pathogens, and while these hens did not necessarily show symptoms of the diseases, they produced immunoglobulins against those pathogens. In such early times, farmers often fed eggs to ill and weak animals, including piglets and young ruminants, to strengthen their immune system – quite often with great success.

Immunoglobulins from milk products

We have been formulating piglet feeds with immunoglobulins since we first started using dairy products, and specifically skim milk and whey powder. A good-quality whey contains about 3 to 4% immunoglobulins (IgA), which although of a weak "potency", still make their presence felt when whey is added at 20 to 30% in a piglet feed. Although IgA are an important source of immunoglobulins, their role is not as specific as that of IgG and IgY in that their actions are more generic in protecting the gut epithelium. In addition, in many cases of over-heated (low-cost) dairy products, immunoglobulins may be destroyed, like any other protein.

Of course, in addition to normal milk, bovine colostrum is a rich source of immunoglobulins (Table 3.1). Indeed, there has been some evidence that providing such ingredient in powder

Table 3.1 *Immunoglobulins in sow's colostrum and mature milk*

Component	Colostrum	Mature milk
Protein (%, N × 6.25)	15	5.5
Casein (%)	1.5	2.75
Whey (%)	15	2
IgA (mg/ml)	21	5
IgG (mg/ml)	96	1
IgM (mg/ml)	9	1.5

Mavromichalis (2006), Applied Nutrition for Young Pigs, *CABI.*

form to piglets may increase post-weaning growth performance. The real problem with bovine colostrum is its very expensive price (over 30 euros per kg) relative to its required high inclusion rate (10 to 20 kg per tonne feed). In addition, there is usually a limited availability of bovine colostrum in the market, and what limited quantities exist, they are usually reserved for calf milk replacers and for orphan animals of all mammalian farm species.

Immunoglobulins from animal plasma

A much richer and more potent source of immunoglobulins than milk and colostrum is, of course, that of animal plasma. This ingredient contains 8 to 25% immunoglobulins (IgG) and its presence in piglet feed at levels of around 30 to 70 kg/tonne has been shown to dramatically increase piglet performance, especially during the first couple of weeks post-weaning.

The original work demonstrating that animal plasma exerts its beneficial effect through its immunoglobulin fraction was conducted at the University of Iowa (Table 3.2). When plasma was fractioned into its component parts, only the fraction containing the immunoglobulins enhanced piglet performance to levels comparable to whole animal plasma. Interestingly,

Table 3.2 *Fractions of animal plasma and their effects on piglet performance post-weaning (grams/day)*

	Casein	Plasma	Albumin	IgG	Rest
Weight gain	19	**134**	78	**158**	50
Feed intake	181	**262**	244	**273**	191

Gatnau et al. (1995), Journal of Animal Science 73 *(Supplement 1):82.*

casein, despite being a prime protein source, did not elicit any response. This helps illustrate the fact that animal protein quality alone in animal plasma is not sufficient to explain its beneficial effects.

These findings were later verified in many other studies by different research groups and have been published in scientific journals. In the last such report in the *Journal of Animal Science* by Pierce et al. (2005) we can read:

> *porcine and bovine plasma are beneficial to young pig performance during the first week after weaning and that the immunoglobulins fraction of plasma is the component that is responsible for the enhancement in growth rate and feed intake.*

Like most ingredients, however, animal plasma is not without its own problems. First, it must be pointed out that although animal plasma contains a relatively high level of immunoglobulins, these are "random" immunoglobulins. This is due to the fact that animal plasma is derived from blood at the slaughterhouse from a wide range of animals raised under different conditions. As such, its immunoglobulin concentration is hugely variable and consequently of extremely unpredictable composition. As the potency of immunoglobulins depends on the actual health of the animals slaughtered and their exposure to different pathogens, it is entirely possible for one batch of plasma to contain 15% immunoglobulins and another as little as 8%, yet there is no guarantee the richer animal plasma will be more effective, as it may contain a less effective profile of immunoglobulins. As such, the level of immunoglobulins can be entirely irrelevant, whereas their specificity to piglet pathogens is always of paramount importance.

Nevertheless, animal plasma maintains a powerful position in modern piglet diets. On the other hand, animal plasma remains extremely expensive, exceeding at various times 4 euros per kg. Assuming a typical commercial prestarter diet with a ceiling of 600 euros per tonne for ingredient cost, it is quickly evident that animal plasma added at the usual rate (5%) will require 200 euros or 1/3 of the total ingredient cost. This allows very little room for other specialty ingredients and it is perhaps the reason why some modern formulas either contain very little plasma, or very little in terms of lactose or cooked cereals, etc. One more reason that animal plasma remains so expensive, perhaps, is the high demand from the pet food industry and lately as an ingredient in broiler super prestarter feeds.

Immunoglobulins from eggs

For many years now, a more sophisticated source of immunoglobulins has existed in the market, in the form of egg immunoglobulins. These are derived from eggs produced by hens challenged or vaccinated against specific piglet pathogens. As such, these eggs are extremely rich in immunoglobulins with a high specificity (both avidity and affinity) for piglet pathogens, such as Escherichia coli for example. As such, lower levels of immunoglobulins are required to elicit a positive response. Obviously, this is a modern approach to feeding farm-raised eggs to sick piglets, as mentioned above.

In addition, research has demonstrated that piglets challenged with various pathogens and then fed with egg immunoglobulins recovered faster and with less loss of performance than control piglets. More importantly, mortality due to disease was reduced drastically when egg immunoglobulins were fed, which

Table 3.3 *Efficacy of egg immunoglobulins against virus-specific diarrhea in newborn piglets*

	Days with diarrhea	Weight gain (%)	Mortality
Control	6.5	**−18.5**	100
1x dose IgY	3.3	**−7.9**	25
2x dose IgY	2.3	**+3.9**	0
4x dose IgY	1.3	**+7.8**	0

Ohsugi et al. (2000), Ghen Corporation, unpublished data.

is indeed of huge economical significance under practical conditions (Table 3.3).

More recently, research and empirical evidence has provided evidence that egg immunoglobulins are also effective in relatively healthy animals (Table 3.4). The term "relatively" is used because really healthy animals do not benefit from in-feed immunoglobulins. The data presented in Tables 3.3 and 3.4 provided sufficient, albeit preliminary evidence that egg immunoglobulins can replace animal plasma. Such application provides a less expensive alternative to animal plasma, or an option where animal plasma is not allowed or not wanted.

Yet, even this new source of immunoglobulins is not without its own problems. First, the immunoglobulin profile is important: obviously, the wider the coverage the better, but it also must be targeted to regional requirements. As it might be expected, not all such commercial products are created equally. Of course, plain egg powder is not the same as that derived from challenged or vaccinated hens, and therefore the two products should not be compared in terms of price or effectiveness. Perhaps, the most important factor determining quality is the way these immunoglobulins are produced.

Table 3.4 *Egg immunoglobulins replacing animal plasma under commercial conditions in diets for weaned pigs*

ADG, g/day	Negative (soy protein)	Animal plasma	Egg immunoglobulins
Trial 1			
Phase 1 (7 days)	137[a]	208[b]	195[b]
Phase 2 (14 days)	477[a]	459[a]	510[b]
Trial 2			
Phase 1 (7 days)	97[a]	102[a]	121[b]
Phase 2 (14 days)	454	449	452
Trial 3			
Phase 1 (7 days)	28[a]	71[b]	93[b]
Phase 2 (14 days)	353[a]	478[b]	437[b]

Means within the same row with different superscript are different ($P<0.05$).

Mavromichalis (2010), unpublished data.

Production methods. Egg immunoglobulins are produced in two ways. In the first, hens are vaccinated with a cocktail of pathogens, their eggs are then collected and processed, giving thus a mixed product of known profile, but a rather impossible to control level of each immunoglobulin. This is because each hen does not respond to the pathogen cocktail in the same manner, producing thus in turn a variable cocktail of immunoglobulins. The second way of producing egg immunoglobulins is by vaccinating a whole flock of hens with only one specific pathogen. This allows the derived product to be identified and titrated to strict specifications. As such, it can be mixed with other similarly produced immunoglobulins against other pathogens to meet specific requirements at required levels. This second method, albeit more expensive, ensures both a known profile and level of immunoglobulins. In addition, it is possible to design specific products. For example, there is today a commercial product that has been designed to target edema disease in piglets, yet another that works best as animal plasma replacement.

Egg immunoglobulins against edema

Edema, or oedema, disease is a complicated yet quite frequent gastro-intestinal disorder affecting millions of piglets each year. Major symptoms include disorders of the nervous system, puffy eyelids, sometimes diarrhea, lameness, loss of appetite, depressed performance, unthrifty overall condition, and, of course, elevated mortality.

Currently, edema disease is thought to be triggered by certain serotypes of Escherichia coli bacteria. As such, the best advice that could be offered is veterinary intervention and a high level of health status throughout all farm operations. Nevertheless, even under the best veterinary and management care, edema still emerges in certain farms. In these persistent cases, dietary measures often prove beneficial in controlling edema disease, but only when used as part of a more comprehensive program.

Such dietary measures include any and all methods that are employed in controlling coli bacteria that may cause piglet diarrhea. These measures include the following: zinc oxide at 3000 parts per million of zinc; copper sulphate at up to 250 ppm of Cu; low levels of dietary protein (down by up to 4 percentage points); reduced calcium levels; increased dietary fiber level, organic acids at high concentrations (up to 1%), and lately, egg specific immunoglobulins.

Commercial evidence. Using egg immunoglobulins is the latest method of addressing edema disease and it is in fact

Table 3.5 *Commercial trial controlling edema disease with egg-derived specific immunoglobulins*

	Control	Immunoglobulins
Mortality (%)	10.0	5.2
Medicine cost (€/pig)	11.5	7.0
Weight gain (grams/day)	402	442

Mavromichalis (2010), unpublished data.

the only direct dietary intervention that attacks the specific bacteria. This is because these immunoglobulins are obtained from eggs that have been produced by hens vaccinated against strains of piglet-specific strains of Escherichia coli known to trigger edema disease. As such, they are not inexpensive!

Drawing from commercial evidence, the use of edema-specific egg immunoglobulins was tested in a commercial farm with strong problems of edema disease. Results (Table 3.5) suggested that pigs fed with egg immunoglobulins specific against edema disease were about 2.5 kg heavier at 68 days post-weaning, while mortality was just above 5% compared to 10% in piglets fed diets without this supplement. Furthermore, treatment cost for medicines was significantly reduced. More work is needed to verify such commercial observations, but the background work is indeed promising.

New nutrient specification: Immunoglobulins or Ig

It has been advocated that immunoglobulins should be labeled as a new nutrient in piglet diets. We still do not know the exact levels required and the exact form that we need to supply them in piglet formulas. To include immunoglobulins as a nutrient into the formulation matrix, it will be important to use a standardized source of such specific immunoglobulins. To this end, a new field of research should be opened with exciting possibilities. In brief, we need to move beyond the qualitative approach and quantify animal requirements and ingredient contributions so that we can automate immunoglobulins in feed formulation and design. Currently, there is much that needs to be done towards this goal.

Personal experiences

As a field nutritionist, I have always advocated the use of immunoglobulins in piglet diets. I basically started my career by using animal plasma to convert very expensive European-type diets (rich in expensive skim milk) into less expensive diets by relying heavily on this ingredient as the driving force for feed intake. When an outbreak of hemolytic Escherichia coli was finally controlled only by the use of egg immunoglobulins (in diets that also contained animal plasma), I became interested in the role of egg immunoglobulins as a functional component to enhance gut health in sensitive farms.

Much later, due to animal plasma being so expensive that it was nearly impossible to use under the conditions I was operating, I was convinced by my own commercial trials that egg immunoglobulins could totally replace animal plasma. I switched gradually to this more sophisticated product until I was able to retain plasma only where local market conditions required its presence, and that in limited amounts.

As a consultant, I insist on immunoglobulins being present even in the least expensive diets, preferring to remove other ingredients. I have designed diets containing immunoglobulins from all different sources, with equal success, and I can attest to their effectiveness. I am not an advocate of the notion that animal plasma retains a level of health risk, as most suppliers today do an excellent job offering high-quality products. My only complaint with plasma has always been its high price!

Notes on immunoglobulins

CHAPTER 4

Replacing zinc oxide in piglet feed formulas

Zinc oxide (ZnO) had been a traditional remedy for piglet diarrheas long before it became public knowledge. It was in 1989 that Dr. H. D. Poulsen announced at the 40th annual meeting of the European Association for Animal Production in Dublin that zinc oxide at pharmacological doses could stop diarrheas in weaned pigs. Following this announcement, many universities followed with their own data trying to verify the diarrhea-controlling properties of zinc oxide. Among the many successful trials, it merits to mention the first negative study at Kansas State University (Table 4.1), only to be followed up by more trials later on that verified what Dr. Poulsen and

Table 4.1 *An early failed attempt to prove the effectiveness of zinc oxide*

	Trial 1		Trial 2	
	Control	ZnO	Control	ZnO
Weight gain, g/day	115	115	322	317
Feed intake, g/day	294	299	494	507

There were no treatment effects (P<0.05).

Kansas State University (1992), Swine Day, p. 90.

others had already discovered: that although zinc oxide is a potent additive, like any other additive including antibiotics, zinc oxide cannot be 100% effective. Nevertheless, zinc oxide soon became a universal additive in controlling post-weaning diarrheas, not least because it was reliable, relatively cheap, and readily available. It is important to note that all these discoveries were done in an era when antibiotics were still used extensively in piglet diets. This can be interpreted as zinc oxide being effective on top of antibiotics, and (or) that zinc oxide's mode of action is not (only) just bacteriocidal.

Another positive note regarding the use of zinc oxide came in 1993, when Dr. D. H. Baker and his PhD student J. D. Hahn, from the University of Illinois, published, again for the first time in the scientific annals, what field nutritionists and producers were already experiencing: not only did zinc oxide prevent or cure post-weaning diarrheas, but it also made healthy pigs grow faster and more efficiently (Table 4.2). Since then, and until measures to restrict its use in several countries were effected, zinc oxide has been considered an indispensable ingredient in almost all types of commercial or home-made piglet diets.

Table 4.2 *First evidence that ZnO improves growth performance in piglets*

Treatments	Weight gain, g/day	Feed intake, g/day
Control	445[ab]	678[a]
3000 ppm zinc from ZnO	500[c]	765[b]
3000 ppm zinc from $ZnSO_4$	414[a]	635[a]
3000 ppm zinc from organic Zn	437[ab]	635[a]

Means within the same column with different superscripts differ (P<0.05).

Hahn and Baker (1993), Journal of Animal Science, *71:3020.*

Finally, it would not be unreasonable to assume that the success of the recent ban on antibiotics in the European Union was at least partially based on the availability of zinc oxide to control post-weaning diarrheas and promote growth. This was evidenced by the epidemic of diarrhea outbreaks when, in the absence of growth-promoting antibiotics, zinc oxide was banned in several European countries. At the end, zinc oxide was allowed again, but it is still under tight regulation and scrutiny.

Issues with high levels of zinc oxide

Although zinc oxide at nutritional levels (up to 150 ppm) is considered safe, higher levels used for pharmacological purposes (up to 3000 ppm) are considered undesirable, for a number of reasons. As such, replacements are currently under investigation, but most have not been satisfactory. But, in order to find suitable alternatives, we must understand first why zinc oxide at pharmacological levels is considered problematic, and second, how it works once ingested by piglets.

Although zinc oxide promotes health and performance in piglets, zinc (Zn) is a heavy metal and as such, toxic to most living organisms, including pigs. In fact, according to the National Research Council of the USA (1998), feeding pigs with excessive levels of zinc (depending on source) for a prolonged period of time causes symptoms of toxicity. As piglet feeds are normally fortified with around 3000 ppm zinc from zinc oxide, for controlling diarrheas and improving growth performance, it is evident that this is bound to cause problems (normal dietary requirement is about 100 ppm zinc). Indeed, and this is based solely on unpublished but repeated observations, feeding 3000 ppm zinc from zinc oxide for as short a period as four weeks post-weaning often causes a marked depression in feed intake towards the end of this period. This is the case especially among pigs exhibiting high early feed intake. Obviously, zinc toxicity is not just an issue of zinc concentration in the feed but of actual zinc intake versus time.

Nevertheless, the actual reason why zinc oxide is controlled in the European Union is because zinc as a heavy metal tends to accumulate in soil when zinc-rich manure from piggeries is applied to the fields. Thus, because high levels of zinc in soil (and due to runoff in water reservoirs) are considered an environmental pollutant and health hazard, the use of zinc oxide at pharmacological levels has been severely curtailed in the European Union. This has caused not a few problems, especially when piglets are removed from one country to another. Such an example are piglets transferred from Denmark (zinc oxide allowed) to be finished in Germany (zinc oxide not allowed), and when this happens, digestive upsets, and especially edema, become the first order of business for German practicing veterinarians.

More recently, it has been demonstrated that prolonged exposure to high levels of zinc can increase some form of resistance in microbes, such as in the case of methicillin-resistant Staphylococcus aureus (MRSA). Although this is not of an alarming nature (yet), it does bring into focus the issue of metallo-resistant genes in gram-negative bacteria (such as Escherichia coli).

Mode of action(s)

To successfully replace zinc oxide, we must fully understand its mode of action(s), something that eludes us as of this day. Indeed, in 2010, X. L. Li and co-workers proposed in a comprehensive review in the *Journal of Animal Science and Biotechnology* no less than six potential modes of action that could be responsible alone or in any combination for the beneficial effects of zinc oxide in piglets. What is clear from the list below is that zinc oxide acts primarily at and through the intestinal level, having both a local and systemic effect on:

- overall zinc status in weaned piglets,
- intestinal microbe flora regulation,
- intestinal structure and function,
- ion secretion in the intestinal epithelium,
- intestinal immunology, and
- secretion of brain-intestine peptides.

According to the same authors, the effects of zinc oxide can be summarized as follows:

> *zinc oxide appears to regulate the intestinal microflora and decrease the release of histamine to prevent diarrhea. Zinc oxide also increases the expression of intestinal insulin-like growth factor-1 (IGF-1) and the IGF-1 receptor to ameliorate intestinal injury associated with weaning, and mediates secretion of brain-gut peptides to stimulate food intake, to promote the growth of piglets.*

Several recent studies have indicated that zinc oxide at pharmacological doses controls the size of major groups of bacteria in the stomach and proximal small intestine. To this end, lactobacteria and colibacteria are affected the most, with the resulting re-balancing act being to the favor of the animal.

At present, it is unclear how this bacterial modulation is affected and how it benefits the animal, especially since lactobacteria are suppressed whereas colibacteria are enhanced.

Zinc oxide alternatives

Clearly there is a long road before we finally understand how zinc oxide works, but the need for an effective replacement is immediate and pressing. To this end, the following products have been tested as immediate replacements for zinc oxide.

Organic zinc. There is some evidence that lower levels of zinc from organic (often chelated) forms may be as effective as higher levels of zinc from zinc oxide. The data are not consistent at best, and the claims are not supported by the fact that inorganic zinc oxide of low or high zinc bioavailability supports equal performance in young pigs (Tables 4.2 and 4.3). As mentioned in the above review by Dr. Li and co-workers, there is more to zinc oxide than just absorbed zinc levels.

Tetra-basic zinc chloride. This different form of zinc has been tested in piglets and results have been promising

Table 4.3 *It is not zinc availability that confers ZnO its high value*

	Negative control	Low-availability ZnO	High-availability ZnO
Weight gain, g/day	218[a]	276[b]	271[b]
Feed intake, g/day	306	340	321

Means within rows with different superscripts are different (P<0.05).

Mavromichalis et al. (2000), Journal of Animal Science, *78:2896.*

Table 4.4   *Tetra-basic zinc chloride (TBZC) as an alternative to normal*
ZnO

Treatment	Weight gain, g/day	Feed intake, g/day
Control	229	329
3000 ppm Zn from ZnO	280	374
750 ppm Zn from TBZC	235	316
1500 ppm Zn from TBZC	279	361
2250 ppm Zn from TBZC	294	384
3000 ppm Zn from TBZC	285	370

Weight gain: control versus ZnO ($P<0.05$), control versus TBZC ($P<0.05$), TBZC linear
effect ($P<0.05$).

Mavromichalis et al. (2001), Canadian Journal of Animal Science, *81:387.*

(Table 4.4). But, according to studies, pigs require at least 1500
ppm zinc from this source to equal performance obtained by
zinc oxide. As these tests were performed using diets with
antibiotics, it merits observing if comparable results can be
obtained with antibiotic-free diets. Nevertheless, even with
half the normal dosage, the problems of zinc toxicity and
accumulation remain unresolved, albeit at a reduced state.
Finally, it should be mentioned that this is the only form of
zinc, other than zinc oxide, that has shown such promising
results in replacing normal zinc oxide.

Zinc sulphate. As can be seen in Table 4.2, adding 3000 ppm
zinc from zinc sulphate failed to elicit any positive response
over the negative control diet, whereas the same amount of
zinc from zinc oxide greatly improved growth performance.
This is evidence that not all salts of zinc are suitable in replacing
zinc oxide. It also points to the fact that it is not the high level
of zinc *per se* that brings about the beneficial effects of zinc
oxide.

57

Encapsulated zinc oxide. There is at least more than one source of fat-covered zinc oxide with claims of comparable results at lower dosages. As zinc absorption does not appear to be the main issue here, as discussed above, and as encapsulation only delays eventual absorption, it is difficult to place this claim in a biological context. Perhaps, the encapsulation prevents zinc oxide from being dissolved in the acidic stomach pH. In fact, it is believed that it is the insoluble nature of zinc oxide (it is insoluble at the neutral pH of the intestinal environment) that confers part of its beneficial effects. But, stomach pH in weaned pigs is nowhere as acidic as in older pigs, and as such we cannot accept that much normal zinc oxide is dissolved to begin with. But, all these are still hypotheses that merit further investigation as, until now, there has been no convincing in-vivo evidence that zinc oxide precipitates in the stomach when it comes in contact with hydrochloric acid.

Porus zinc oxide. The effective surface of minerals, with which they can contact microbes to exert their bacteriostatic and bacteriocidic effects, can be enhanced by methods other than grinding. For example, this can be accomplished through absorption by clay minerals (a method still in its infancy) or by creating a porous structure. The latter is another promising idea with much commercial interest. Not only standardization is possible, but the final product can achieve an enhanced surface area and more importantly greater porosity as much as 10 times higher than normal zinc oxide. Published results (Table 4.5) indicated that it replaces normal zinc oxide with good results, especially after the second week post-weaning.

Table 4.5 *Porous ZnO can replace high levels of normal ZnO*

	No ZnO	3000 ppm Zn normal ZnO	300 ppm porous ZnO
Weight gain, g/day	283[a]	310[b]	318[b]
Feed intake, g/day	371	373	380

Means within the same row with different superscript are different (P<0.05).

Cho et al. (2015), Animal Science Journal, *86.617.*

Nano-zinc. It can be theorized that finely ground zinc oxide can be used at regulation levels (150 ppm in the final feed in the European Union) and still ensure pigs grow better and without diarrheas. The theory is based on the hypothesis that a finely ground zinc oxide product will expose more molecules of zinc oxide to interact with the gastro-intestinal tissues and microbial population. For example, normal zinc oxide has a surface of four square meters per gram, whereas with nano-grinding, this number can be increased to a high, albeit very variable, level. On the negative side, such sources arrive as industrial by-products that can be highly contaminated with other heavy metals. Plus, they tend to be quite variable in quality, and certainly, there is a dearth of scientific information on their use as feed additives, apart from some commercial trials.

The future of zinc in pig feeds under EU regulations

In 2003, the European Union replaced the upper limit of zinc concentration in pig feeds from 250 ppm (mg/kg complete feed as-is basis) down to 150 ppm. As most complete feeds contain 30 to 40 ppm natural zinc, this allows supplementation, mostly from zinc oxide or zinc sulphate, at no more than 110 ppm zinc. For most pig feeds, this level is more than enough as requirements are no more than 100 ppm. For weaned pigs, there is a problem with this upper limit. With the EU regulation 1334/2003 the widespread use of zinc oxide at pharmacological doses (from 2000 up to 4000 ppm zinc) stopped, only to be re-allowed under veterinary prescription in 2005 to confront the problem of poor gut health as growth promoting antibiotics were phased out.

Why is zinc regulated? As mentioned above, zinc is toxic to living organisms (animals, bacteria, plants) when encountered in high concentrations. Pigs are affected at levels above 1000 ppm zinc for more than four weeks. Plants suffer when soil concentration in zinc exceeds 100 ppm, whereas bacteria in soil appear a bit more resistant at 150 ppm. Although animal toxicity can happen only very rarely, this is not the major concern of the EU regulation agencies. Instead chlorosis is the main concern, as this might curb agricultural productivity in EU – after all, pig manure must be spread in arable land.

What is chlorosis? Plants require iron as part of their chlorophyll photosynthesis system. They acquire iron from the soil, and there is rarely a deficiency of iron. But, heavy accumulation of trace minerals such as copper (another EU regulated mineral) and zinc can bind such iron in the soil, making it unavailable to plants. Not only that, but zinc competes with iron absorption sites in the plant, further promoting this secondary iron deficiency state. Thus, plants cannot synthesize chlorophyll and appear pale (hence, the term chlorosis). In turn, the lack of chlorophyll reduces photosynthesis, leading to lower productivity. In other words, affected plants, cereal crops for example, produce less in heavily contaminated fields.

It has been estimated that between 1973 and 1983, zinc accumulation in Western Europe has been about 0.41 ppm per year. Given the present burdened state of most arable land in the EU, it is not unreasonable to expect such lands to reach toxic levels within a few generations from now. Some pessimistic estimates call for 80 years until such premium arable lands become useless. And this is why the EU is rethinking its zinc (and copper) policy.

Zinc and phytase. Another aspect of zinc in animal nutrition is the interaction of zinc with phytase. This enzyme is capable of releasing zinc bound (and hence, unavailable to the animal) in phytate, along with phosphorus. The EU is thinking that the widespread use of phytase may allow for a further reduction on the upper limit for zinc. For example, it has been estimated that using phytase may provide an incentive to reduce the zinc upper limit from 150 down to 110 ppm, which would allow only up to 70 ppm zinc to be supplied by inorganic forms.

Determining zinc needs. Perhaps the most important issue that needs to be addressed before even discussing upper limits for zinc is the establishing of accurate requirements. Currently, the accepted zinc requirement levels are based on old(er) data derived from young(er) pigs. Thus, zinc requirements for older pigs are extrapolated from such experiments. Plus, the requirements are based on "total" zinc, whereas a more accurate expression would be on a "digestible" basis – not unlike the case for amino acids and phosphorus. But, such trials require centralized organization and funding, and here is where the combined resources of the European Union can come into play for the benefit of the community.

Bioavailable forms of zinc. Naturally, the use of highly bioavailable forms of zinc, such as chelated sources, can provide an alternative to minimize zinc excretion in manure and subsequently in soils. But, and this is a very big "but", before we employ such forms of zinc: (1) scientific proofs must be established (surprisingly, all EFSA Opinions published in the last few years failed to conclude that all chelated forms had higher bioavailability); and (2) we must establish how much we can lower zinc dietary specifications (this question intersects the need of establishing requirements on a digestible basis.) As it stands, bioavailability is a relative term that has specific limits towards reducing zinc concentrations in feed and manure. Nevertheless, empirical calculations can be relied upon, at least to limit the use of generous safety margins.

Weaned versus older pigs. Here, we must consider two alternatives. One is the approach discussed in Belgium: that is, to reduce zinc in diets for growing-finishing pigs in exchange of allowing pharmacological dosages in weaned pig diets. In other words, instead of looking at zinc in feed, we can focus on total farm zinc output. This works best when farms are farrowing-to-finish, as all-nursery facilities would have trouble following this scheme. The other approach is to look for alternative forms of zinc oxide. That is, zinc oxide products that can be used at low(er) concentrations but with comparable effects such as those observed with normal zinc oxide used at pharmacological dosages. For nutritional purposes, some organic zinc products are definitely more bioavailable, and as such they can be used to lower zinc excretion from all pigs, not only weaned ones. This can contribute towards a greater allowance for zinc oxide in post-weaning piglet feeds. For pharmacological purposes, the alternative forms discussed above offer considerable ground for exploration, but current upper limits must be reconsidered even for these products.

Clearly, zinc is a problem area for the EU – and predictably for other pig-producing regions – and as such we should anticipate tighter control over its use and disposal. Being proactive, it is recommended to focus first on establishing accurate animal zinc requirements, on a digestible basis. Then we must continue exploring the use of more "friendly" forms of zinc, such as highly bioavailable zinc to cover nutritional needs, and protected or porous zinc oxide to protect gut health in weaned pigs. All require generous funding from EU and the private sector.

Zinc oxide buffers organic acids in piglet feeds

It is not widely known that zinc oxide has a strong buffering capacity, reducing thus the acidity in the stomach of newly weaned pigs. Perhaps this is not as important as in the case of antibiotic-free formulas. Nevertheless, it is worth mentioning that the usual 2 to 3 kg zinc oxide per metric tonne of complete feed neutralizes the equivalent of most organic acids. For example (http://animine.eu/ABC4-calculator/), examining two different diets, one with normal zinc oxide and another with a low-inclusion zinc oxide alternative, the following remarks can be offered for consideration:

- Adding 0.1% citric acid reduces the acid-buffering capacity of the feed by 56 mEq/kg.
- Adding 0.03% zinc oxide increases the acid-buffering capacity by 48 mEq/kg.
- For reference, the recommended acid-buffering capacity at pH 4 is about 350 mEq/kg.

As it appears, with normal zinc oxide and when the feed is high in acid-binding capacity, organic acids must be added to control the effect of zinc oxide, or an alternative must be sought. It is obvious, low levels of organic acids (1 to 3 kg/ MT) are practically pointless in diets containing normal zinc oxide as their acidity is neutralized – unless they are used for this purpose alone, but this greatly increases the cost of using normal zinc oxide.

In reverse logic, when replacing normal zinc oxide with alternative forms at lower doses, then organic acid inclusion levels can be reduced, offering the potential for considerable feed cost savings. Or, the extra stomach acidity might be counted on to use other ingredients that require a lower pH, such as higher levels of soybean meal. In diets with an already low acid-binding capacity, alternative forms of zinc oxide will allow high levels of organic acids to provide a very strong bacteriostatic and bactericidal effect, as predicted by their own characteristics.

Perhaps, the above conclusions may at least partially explain the current commercial situation in the USA and the European Union countries. In the USA and other countries where normal zinc oxide is included at about 3 kg per metric tonne, there is now increasing evidence that some piglet diets benefit from organic acids. In Europe and other countries where in-feed antibiotics are banned, previous experience has shown that the use of low levels of organic acids is pointless. Was this because up to now most such diets contained, under previous legislation, high levels of normal zinc oxide? If this holds true, should we expect now to derive a greater benefit from low inclusion rates of organic acids in diets with alternative forms of zinc oxide?

Personal experiences

A third of my PhD thesis at the University of Illinois was on zinc oxide with two published research papers. Thus, I had a first-hand experience on this additive, and most importantly in one potential replacement. As a field nutritionist in the USA, I had always used and counted on the benefits of zinc oxide, but in my work in Europe, zinc oxide was something that caused more frustration than not. Not only it was first banned, then re-allowed, and now it is tightly controlled, but also, such measures are not followed universally throughout the European Union.

Thus, as a nutritionist, I cannot depend on normal zinc oxide for products that must be acceptable in different markets with different legislations over the same problem. As such, I have explored alternative forms of zinc oxide for such cases. I believe I have found the right replacement strategy, but its use must be part of a well-thought reformulation effort. In particular, I believe the use of such alternative forms of zinc oxide must be accompanied by an effective blend of organic acids – and low crude protein levels – to provide the best protection against most bacterial pathogens.

Finally, I encourage all my clients to prepare for the eventual removal of normal zinc oxide from our list of available feed ingredients. The problem of soil accumulation and consequent plant chlorosis is real, and if unchecked, it is bound to affect the very next few generations, limiting food supply. I expect most countries to continue pressuring for reduced zinc excretion in the environment, and for that I believe it is best to be proactive, even if zinc oxide is still allowed at high levels.

Notes on zinc oxide

CHAPTER 5

Replacing lactose in high-value piglet formulas

Lactose is a choice energy source for piglets. Chemically, it is a disaccharide composed of galactose and glucose (as opposed to sucrose, which is a similar disaccharide composed of two identical glucose molecules). Lactose is digested by the enzyme lactase, which abounds in newborn piglets, but its activity diminishes with age.

In most commercial piglet feeds, lactose plays a central role determining (a) final price and (b) quality, in terms of piglet feed intake. Lactose sources are invariably expensive, compared to other sources of energy such as starch and lipids. The most common sources of lactose are skim milk (used extensively in Europe) and dried whey (worldwide), with about 50 and 70% lactose, respectively. Other dairy products that may be less expensive yet equally effective in supplying lactose include crystalline lactose, deproteinized whey, whey permeate, milk chocolate product, cheese powder, yogurt powder, and whey protein concentrate or isolate. Quality in all dairy products is more important than lactose and (or) protein concentration.

Increased use of milk products to replace other animal-derived foods for humans continues to drive up demand and price for all dairy products. As such, the inclusion of lactose in piglet feeds remains an expensive investment, especially for regions that are net importers of dairy products. To this end, there is a perpetual interest in ingredients that can replace lactose without sacrificing feed quality and consequently piglet performance and health. This chapter reviews lactose alternatives and discusses the concept of lactose equivalents in feed formulation.

Why piglets require lactose after weaning

To replace lactose, it is imperative to understand why piglets require this simple sugar to thrive after weaning. First, it should be noted that pigs less than eight to ten weeks old cannot utilize starch very effectively. Although the digestive system of pigs as young as three weeks of age can be induced rapidly to utilize raw starch within days, weaning stress accompanied by low feed intake prolong physiological maturation of the digestive system. For this reason, simple sugars such as lactose and gelatinized starch (cooked cereals) are generally more digestible than raw cereals for the first two to three weeks post-weaning. In other words, weaned pigs require readily digestible sugars to thrive until their digestive system is fully ready to digest raw starch.

In practical terms, when diets based on cereals and plant protein sources are supplemented with dairy products, such as dried whey and skim milk, growth performance of young pigs is markedly improved. Early research and empirical evidence from the USA has shown that inclusion of 10 to 20% dried whey in simple diets (corn, soybean meal, and oat groats) consistently improved growth performance by at least 15% in pigs weaned around three weeks of age. Other dairy products, especially skim milk, are at least equally effective in enhancing post-weaning growth performance, although they are more expensive than dried whey and its co-products. In Europe, and especially in Great Britain, skim milk is valued even more than whey, even though it is always more expensive.

It is interesting to note the effectiveness of dairy products was attributed initially to the observed increased palatability and the higher digestibility of milk protein. However, it has been clearly demonstrated by now that lactose, and not protein, is responsible for improved feed intake and weight gain, especially during the first week post-weaning. A classic study conducted at The Ohio State University indicated that lactose supplementation restored growth performance in diets devoid of dried whey, whereas adding only lactalbumin failed to do so. Yet, a highly digestible source of protein (such as feed–grade amino acids) but not lactalbumin necessarily, was essential for lactose to elicit its beneficial effects in whey-free diets (Table 5.1).

In the same study, it was also demonstrated that after the first two weeks post-weaning, piglets respond to lactalbumin

Table 5.1   *Effects of whey, lactose, and lactalbumin on growth performance in young pigs fed simple corn-soybean meal diets*[1]

	A[2]	B	C	D	E
Weight gain, grams/day	210[a]	233[b]	208[a]	251[b]	243[b]
Feed intake, grams/day	341[a]	395[b]	371[a]	398[b]	388[b]
Gain:feed, grams/kg	616[a]	590[b]	561[b]	631[a]	626[a]

[1] A total of 420 pigs (6.8 kg) in 14 replicates per treatment were used in a 21-day growth assay. Means within the same row with different superscript are different (P<0.05).
[2] A = corn and soybean meal diet (negative control), B = as A plus 20% dried whey (positive control), C = as A plus lactalbumin and starch, D = as A plus lactose and feed–grade amino acids, E = as A plus lactalbumin and lactose.

Mahan (1992), Journal of Animal Science, 70:2182.

supplementation alone, but not to lactose. This is an indication of a rapidly maturing digestive system and the continuing need for a highly digestible protein source even after the initial phase.

How much lactose is needed in post-weaning feeds

One method to control piglet feed cost is to prevent overfeeding of lactose. To achieve this feed formulation goal, accurate estimates of lactose requirements are needed according to body weight and overall diet composition. It is also important to have a feed cost maximum so that lactose specifications can be adjusted accordingly to market requirements.

Several modern studies have investigated minimum dietary lactose concentrations that promote maximum growth performance in weaned pigs. These data suggest dietary lactose concentrations can be rapidly lowered after the initial two weeks post-weaning, whereas pigs over 10 kg in body weight rarely benefit from lactose, especially if they have enjoyed high feed intake levels prior to achieving this body weight. For example, recent research from The Ohio State University indicated that levels up to 30, 20, and 15% were necessary to maximize weight gain in pigs up to 7.0, 12.5, and 25.0 kg body weight respectively. The latter was significantly evident only for pigs lighter-than-age, as heavy pigs did not benefit from lactose.

Although responses to lactose have been somewhat variable due to basal diet composition, environment, and health status of the animals, most experiments tend to point to a similar direction regarding lactose requirements for weaned pigs. In essence, the more the better as long as the extra lactose

Table 5.2 *Recommended lactose concentrations in piglet feeds*

Body weight (kg)	Minimum[1] (%)	Optimum[2] (%)	Maximum[3] (%)
Sow's milk[4]		25	
Less than 4	20	25	30
4–6	15	20	25
6–8	10	15	20
8–10	5	10	15
Over 10	0	5	10

[1] For use in low-cost production systems and antibiotic-free programs.
[2] A balance between ingredient cost and piglet performance.
[3] To accelerate piglet performance, especially under heavy antibiotic use.
[4] On a 90% dry matter basis, which is equivalent to a typical piglet feed.[?] For other weight ranges use average of adjacent values.

does not act as a laxative. In practical terms, combining research and field experiences, it is possible to create a set of commercial dietary lactose specifications such as in Table 5.2. These figures have yielded consistent results in rather disparate commercial conditions, but they should be used as a guiding principle only.

Preliminary research has indicated that in-feed immunoglobulins (such as from animal plasma and eggs) may warrant lower lactose specifications. This might be because immunoglobulins promote high gut health, which in turn drives up feed intake and speeds up digestive maturation. In a US study, pigs fed diets with 6.75% animal plasma required only 15% dietary lactose for maximal growth performance during the first week post-weaning. In contrast, when animal plasma was replaced with soy protein concentrate, performance peaked between 30 and 45% lactose. It appears that when feed intake is already high (due to other feed ingredients or simply because of high health status), lower levels of dietary lactose might be enough. These preliminary results, however, require further investigation and verification, but empirical evidence already points toward this direction.

The value of good-quality whey

Whey is perhaps the most widely used source of lactose in piglet diets. However, whey is not only lactose. Recently, it has been demonstrated that spray-dried whey may inhibit binding of pathogenic Escherichia coli to intestinal mucosal surface and thus prevent infection. Casein, the protein in skim milk, has also been shown to have immune-stimulating effects. Thus, discarding protein quality in dairy products is not completely right under current scientific thinking. In brief, it is important to consider the whole ingredient as a source of valuable components. And, proteins are the most vulnerable because overheating can easily destroy them.

If the mode of action of such common ingredients as whey and skim milk are still not fully appreciated, the effects of their variable quality are even less understood. Skim milk quality is usually from acceptable to exceptional as most batches arrive from the human food industry. Whey quality, however, is more variable not only between sources but even within batches from the same cheese factory. As early as the 1950s, the research community was aware that roller-dried whey was inferior to spray-dried whey, and that "browned" or over-processed whey was generally of inferior quality (Table 5.3).

Quality is a very subjective aspect, and includes, but is not limited to, the following variables.

Table 5.3 *Overheated whey reduces piglet performance*

Whey (20% in feed)	Feed intake preference (%)	
	2–5 weeks	5–6 weeks
Normal	95	71
Browned	5	29

Catron (1960), Proceedings of Distillers Feed Conference, *15:60.*

Chemical analysis. Although it has been repeatedly demonstrated that chemical analysis is a very weak index of whey quality, as far as piglet performance is concerned, adherence to certain specifications allows at least for some stability in formulation. Nutrient specifications for high-quality whey follow the values given in Table 5.4, but products from a specific cheese factory can vary according to the processing method followed and in this case consistency is more important. A vigorous quality-control program is required when large batches are purchased from a great number of sources.

Table 5.4 *Typical chemical specifications for sweet whey*

Item	Unit	Value	SD
Dry matter	%	96.40	
Crude protein (N × 6.25)	%	12.60	
Lipids	%	2.10	2.10
Ash	%	8.70	2.30
Lactose	%	72.9	0.90
Calcium	%	0.82	2.10
Phosphorus	%	0.69	
Sodium	%	0.64	0.15
Chlorine	%	1.76	0.07
Lysine	%	0.94	
Methionine	%	0.20	

INRA (2002), Tables of composition and nutritional value of feed materials, Paris.

Ash and minerals. It should be noted that ash and specifically sulfur and sodium are components highly undesirable in whey as they are linked to osmotic diarrhea, albeit actual research is limited in piglets to support such notion. German researchers investigated the effect of high sulfate concentration in 29 samples of whey on fecal consistency in pre-ruminant calves. Sulfur content in whey ranged from 0.03 to 4.30% per kg dry matter. As expected, a high concentration of sulfate resulted in marked increases in incidents of diarrhea. Extrapolation of these data could lead us to adopt a safe level of sulfate in complete piglet feeds of below 0.25% on an as-fed basis.

Sweet versus acid whey. This is an area where research is rather scant and inconclusive. Nevertheless, commercial experience indicates that high-quality sweet whey is often superior to acid whey of similar specifications. This might be the result of increased palatability and reduced mineral content in sweet whey. In any case, superior-quality acid whey is always preferable to low-quality sweet whey, but this is based on empirical evidence.

Coloration. High-quality whey should have an off-white creamy color. Certain batches may be more yellow in color, due to processing and (or) high levels of natural pigments in milk; so this is normal. Nevertheless, a brownish color is always indicative of excessive thermal processing and formation of Maillard reaction compounds, which diminish the nutritive value of whey (reduce lysine availability) and may also affect its palatability (bitter compounds). This is why spray drying is considered superior compared to other processing methods.

Simple sugars that can replace lactose

Although piglets require lactose over starch to thrive, studies contacted from as early as the 1950s indicated piglets can also use other forms of simple sugars. For example, readily available monosaccharides (glucose, fructose, maltose), and oligosaccharides (sucrose, maltodextrins) are as effective as lactose in supporting feed intake and growth performance. Because of this, the term "lactose equivalent – LacEq" is therefore more appropriate to express dietary specifications for simple sugars, as piglets do not appear to have a requirement for lactose *per se*. Below is a list of sugars that can replace lactose.

Sucrose. Increasing diet sweetness by adding sucrose (common table sugar) has long been practiced in the piglet feed industry. Yet, in complex diets, which are inherently more palatable and digestible, as much as 20% added sucrose does not enhance palatability compared to similar levels of lactose. It is known that young pigs show a strong preference for sucrose only when they have free-choice access to sucrose-based diets, but not when they are offered only one diet. Although responses to sucrose as related to increased palatability are expected to be less pronounced with complex diets, many commercial formulations still contain sucrose or artificial sweeteners.

The use of sucrose to replace lactose in complex diets was investigated at Kansas State University by Dr J. D. Hancock and his graduate students using weaned pigs (6.4 kg) during a 30-day growth assay. Pigs were fed simple or complex diets that were supplemented with lactose or sucrose (20 and 14% during days 0 to 10 and days 10 to 30, respectively). The basal diets contained 4% plasma protein and all diets were medicated, including zinc oxide at pharmacological levels. Results indicated that pigs utilized sucrose as efficiently as lactose, without suffering from osmotic diarrhea.

Earlier work with sucrose-based diets indicated diarrhea in very young pigs. It is now known that diarrhea in neonatal pigs from excessive sucrose consumption is due to the low activity of intestinal sucrase (the enzyme that breaks down sucrose). Sucrase activity increases rapidly with age and by the end of the second week of age, pigs are able to fully utilize sucrose. Therefore, it is recommended that high levels of sucrose not be used in diets for pigs less than 10 days of age.

High concentrations of sucrose may disturb dietary osmolality and cause osmotic diarrhea, but this problem does not occur under most practical conditions, especially when measures are taken to reverse the osmotic effect or add ingredients that absorb and retain the extra moisture.

Molasses. Molasses is a by-product from the extraction of sucrose from sugar beets or sugar cane. It is a viscous semi-liquid with about 75% dry matter and 50% sugars. Sucrose in molasses is in equilibrium with glucose and fructose. Molasses is also rich in ash (9%) and most of its nitrogen is in non-protein compounds (it should not be considered in feed formulation).

High levels of molasses in pig diets are often associated with skepticism regarding digestive disturbances from its high fructose and ash concentration. Molasses is particularly rich in potassium, which may upset the electrolyte balance, increase osmotic pressure and lead to osmotic diarrhea. However, such a hypothesis has yet to be proven. In a Cuban study, minerals included in molasses-free diets at concentrations equal to those provided by high levels of molasses did not cause diarrhea.

Table 5.5   *Molasses versus lactose in complex nursery diets*

Item	Lactose	Lactose and molasses (50:50)	Molasses
Day 0 to 10			
Weight gain, grams/day	297	298	324
Feed intake, grams/day	283	297	317
Gain/feed, grams/kg	1049	1003	1022
Day 10 to 30			
Weight gain, grams/day	473	463	468
Feed intake, grams/day	623	640	655
Gain/feed, grams/kg	759	723	715
Overall			
Weight gain, grams/day	409	404	414
Feed intake, grams/day	504	521	538
Gain/feed, grams/kg	812	775	770

A total of 210 pigs in 8 replicates per treatment. There were no differences among treatments ($P<0.05$).

Mavromichalis et al. (2001), Animal Feed Science and Technology, *93:127.*

A study at Kansas State University with cane molasses indicated that pigs utilized molasses as efficiently as lactose in diets containing as much as 20% molasses during the initial 30 days post-weaning (Table 5.5). In fact, feed intake in pigs fed molasses-based diets was numerically increased (+9%) during the first 10 days post-weaning, indicating a possible taste preference for molasses. In the same study, the youngest pigs were at least two weeks old, and no visual symptoms of diarrhea were observed in daily inspections nor was any pig treated due to impaired health status.

Dextrose. Dextrose monohydrate is derived from starch hydrolysis. It is comprised of crystallized glucose containing one molecule of water for each molecule of glucose and is the primary source of dextrose in food applications, although dextrose anhydrate is also available. Dextrose monohydrate has a 90% lactose equivalent value on a dry matter basis compared to crystalline lactose because of the degree of hydration. Dextrose is highly hydroscopic and tends to cake during prolonged storage – this is important in tropical climates.

Young pigs utilize glucose (dextrose) as effectively as lactose immediately after birth. Thus, dextrose can be used as an alternative to crystalline lactose in liquid milk replacers and highly complex piglet feeds. In a US study, dextrose successfully replaced 12% dietary lactose, in post-weaning piglet diets without animal plasma, whereas in diets with animal plasma, partial replacement of lactose by dextrose at a 2:1 ratio supported at least equal performance compared to all-lactose diets. Diets with two sources of simple sugars were slightly more efficacious than diets with one source of sugars, indicating that young pigs may actually benefit from sugar blends.

Maltodextrins. Maltodextrins are glucose disaccharides derived from starch hydrolysis with a dextrose equivalent (DE) value of less than 20. Dextrose equivalent is a grade specification indicating percentage concentration of reducing sugars (i.e., monosaccharides), an indirect estimate of degree of hydrolysis.

Korean researchers fed weaned pigs complex diets based on lactose or maltodextrin for three weeks post-weaning. No differences in growth performance were observed, indicating that maltodextrin is a suitable source of simple sugars that can totally replace lactose. In the same study, sucrose and dextrose were equally effective as lactose in promoting growth performance, in contrast to raw cornstarch, which failed to improve performance. In a French study, maltodextrin in diets for piglets between 10 and 62 days of age outperformed lactose.

On the negative side, maltodextrins in moderate dietary concentrations (5 to 10%) have been shown to depress palatability of pelleted diets for growing-finishing pigs, possibly due to excessive heating during pelleting. Thus, it is suggested that maltodextrin-based diets be pelleted at even lower temperatures than lactose-based diets. A 60:40 blend of maltodextrin and sucrose depressed growth performance in suckling pigs fed milk replacer, possibly due to fructose toxicity from sucrose. It appears that maltodextrin application in piglet feeds requires some sophistication in formulation and manufacturing.

Corn syrup solids. Corn syrup solids are derived from starch hydrolysis and have dextrose equivalents of 20 or more. Corn syrup solids are used primarily in beverages for humans but substantial quantities often become available for animal feeds.

A US research team used corn syrup to replace 40% crystalline lactose in a liquid milk replacer, offered ad libitum to one-day-old pigs until weaning at 20 days of age. Two grades of corn syrup were evaluated, 20 and 42 DE. Total replacement of lactose with either grade corn syrup did not affect growth performance, digestive development, or gastro-intestinal architecture in neonatal pigs. Piglets grew at 380 grams per day, reaching almost 10 kg body weight at only 20 days of age; an astonishing growth rate illustrating the power of well-formulated (albeit expensive) liquid milk replacer programs.

High-fructose corn syrup. This is another starch derivative, derived from the enzymatic conversion of dextrose to fructose.

As such, it contains between 40 and 90% fructose. In a French study, weaned pigs (21 to 28 days of age) preferred diets with 5% sucrose to diets with saccharin or fructose syrup up to 10% of the diet. Nevertheless, when not given a choice, pigs consumed comparable amounts of each diet. Thus, fructose products may be used in starter formulations if priced competitively to lactose and other simple sugar sources.

Non-sugar lactose alternatives. Inulin, cooked starch, resistant starch, chicory pulp, and many other ingredients have been tested as lactose replacements – all without success. Here it should be noted that lactose probably has a mild prebiotic effect on hind gut microflora. As such, a small part of lactose (perhaps 2%) may be replaced by other ingredients with similar prebiotic effects. Such ingredients are usually sources of functional fibers (inulin, chicory pulp, etc.) However, it is important to understand what is replaced by such ingredients and not confused with the nutritive value of lactose and other sugars.

Concerns with alternative sugars

Replacement of lactose with other simple sugars is not without potential problems. For example, replacing lactose with 5% sucrose or 10% fructose increases pellet friability and softness. Also, reactive (reducing) sugars such as lactose and glucose readily react with amino acids during pelleting in the Maillard reaction. This process renders amino acids, especially lysine, unavailable to the animal. Therefore, lower pelleting temperatures (less than 55°C) and even cold pelleting is often suggested when diets contain high amounts of reducing sugars.

Table 5.6 *Relative sweetness of natural sugars (sucrose is arbitrarily given a relative value of 100)*

Sugar	Sweetness index
Fructose	180
Sucrose	100
Dextrose	70
Lactose	70
Galactose	60
Starch	10

Glacer et al. (2000), Food Chemistry, *68:375.*

Furthermore, as simple sugars differ considerably in sweetness, changes in diet palatability should be taken into consideration (Table 5.6). Therefore, when sucrose or molasses replaces lactose, the use of an artificial sweetener may no longer be warranted. In general, piglets prefer simple sugars in the following order, always when given a choice:

sucrose > fructose > maltose = lactose > glucose > galactose

This order of preference does not take into account the use of modern artificial sweeteners, whereas the use of stevia as a natural alternative sweetener remains rather unexplored.

In addition, excessive concentrations of simple sugars are always possible to predispose pigs to osmotic diarrhea. For instance, by changing the amount of sugars (sucrose and corn syrup solids), dietary osmolality was manipulated between 250 and 700 mOsm/kg. Absorption of water and carbohydrates from the small intestine in piglets increased, and thus chances for diarrhea were minimized, as osmolality increased, at similar dietary electrolyte balance. Maltodextrin increases water absorption, whereas glucose increases water secretion in the gastro-intestinal tract, indicating that sugar oligomers are more effective than monomers in preventing osmotic diarrhea.

Today, it is still largely unknown how gut microflora is affected when lactose is replaced by other simple sugars. It is currently accepted that the trophic effects of lactose on lactobacteria are diminished in diets that contain a plethora of antimicrobial agents and (or) zinc oxide and copper sulfate. The role of lactose and its alternatives in such diets is usually limited to supply a readily source of energy. For that purpose, all lactose alternatives are equally efficacious and replacement can be made based on their energy content and sugar concentration. In antibiotic-free diets, however, lactose should remain the principal, but not necessarily the only, source of simple sugars until further research becomes available.

Lactose equivalents – new formulation term

Ideally, alternative sources of simple sugars should not replace lactose on a weight basis, although this happens more often than not. Nutritionists can use the term lactose equivalent (LacEq) that reflects the concentration of simple sugars in products that can replace lactose. For example, 0.97 kg table sugar (98% sucrose or 98% LacEq) can replace 1 kg crystalline alpha–lactose (95% lactose or 95% LacEq). Molasses, however, contains about 75% dry matter and only 47% sugars. But, it would take 2.02 kg of liquid molasses to replace 1 kg of crystalline lactose. In practical feed formulation, this translates to a new nutrient specification (LacEq) that can be calculated easily for all ingredients that contain simple sugars (Table 5.7).

Table 5.7 *Lactose equivalents (LaqEq) in simple sugars for young pigs*

Product	DM, %	ME, Kcal/ kg	Sugars, %	LacEq, %
Lactose	96	3435	95	95
Dextrose, anhydrous	99	3535	98	98
Dextrose, monohydrate	90	3260	89	89
High-fructose corn syrup	71–77	variable	42–55	42–55
Regular corn syrup	77–85	variable	44	44
Cane or beet molasses	75	2530	47	47
Sucrose	99	3635	98	98
Starch	99	3535	98	0

In a similar manner, lactose requirements can be expressed in LacEq terms, still using the values given in Table 5.2.

Future research needs

It is currently unknown if thermally processed cereals affect the requirement for lactose. It is possible that increased starch digestibility due to thermal processing may provide enough readily available energy to reduce lactose requirements. Although we have some empirical evidence toward this direction, we are lacking scientific proof. In contrast, resistant (retrograded) starch reduces starch digestibility, and thus may actually increase lactose requirements for energy purposes.

It has been speculated that the inability of digestive enzymes to effectively reach the inside of the starch granule is at least partially responsible for the lack of response of young pigs to raw starch. Is it possible for lactose requirements to be lower in diets based on finely ground cereals? And what about the role of starch-digesting external enzymes, such as alpha-amylase?

In contrast, if access to the starch granule is impeding efficient starch digestion, then cereals that increase digesta viscosity may also increase lactose requirements. Thus, diets based on wheat, barley and rye may require more lactose to support performance equal to diets based on corn, sorghum, or rice.

In addition, commercial experience has shown that diets without the usual growth-promoting antibiotics should be formulated with less lactose. This was done as a response to the increased incidents of diarrhea when such antibiotics were first removed. Since then, however, effective antibiotic replacements have been identified but lactose levels remain low in commercial feeds. Is it possible that piglets fed such improved diets with suitable antibiotic alternatives still suffer from low feed intake, due to low lactose specifications, without reason?

Finally, it merits expanding on the preliminary data concerning the use of lactose in diets low or high in immunoglobulins not only from plasma but also from other sources, such as those from eggs. This can have a tremendous impact on feed cost, as all such ingredients are expensive.

Personal experiences

My M.Sc. thesis at Kansas State University was based partially on lactose and its alternatives. Since then, I have formulated countless piglet feeds based on lactose alternatives. Of particular interest is a case in Midwest USA where an inexpensive and abundant supply of cane molasses was used to create a complete piglet nutrition program using molasses exclusively. This was a commercial line of feeds using up to 20% molasses. Of course, pellet quality suffered, and feed bags leaked, but these were problems easily solved. At the end, feed cost was reduced by 25%, whereas feed intake was actually increased. In general, however, I prefer to use a blend of different sugar sources.

As a nutritionist, I have always favored high lactose levels because I can attest to the high feed intake associated with them. But, on the other hand, I have come to realize that a low-quality dairy product can ruin the best piglet feed. Thus, I prefer to use less of a high-quality dairy product, when I need to constrain feed cost, rather than revert to a less expensive dairy product.

I have also witnessed the birth and death of commercial lactose alternatives based on functional fibers, plus different starch products, from a number of sources, and my effort has been to elucidate to users what is what and what is not. I continue to use lactose and its alternatives as a nutrient-energy source, whilst leaving the role of prebiotics to proper functional fiber sources.

From my experiences with antibiotic-free diets in Europe, I can attest to the fact that high lactose levels contribute to soft fecal consistency, which is non-pathogenic. Nevertheless, it is an undesirable phenomenon in some markets and to this effect I prefer to use absorbents instead of reducing lactose specifications. Lamentably, there are markets that prefer a "clean" pig instead of a rapid-growing one, and despite the initial savings in feed cost, such approach always reduces profitability.

Finally, on the issue of skim milk versus whey, I have had good experiences with both and I would rather use them combined if I did not have to choose one over the other based on cost, availability, or market preferences. In general, skim milk performs excellently in diets for pigs less than 4 kg in body weight, whereas its return on investment diminishes as body weight increases. As such, it is best used in milk replacers, creep feeds, and super prestarters. Whey is more than enough for heavier pigs, and where cost is the main target in formulation, whey is my dairy product of choice – always preferring the best possible quality.

Notes on lactose